Digital Audio Signal Processing

Digital Audio Signal Processing

Second Edition

Udo Zölzer
Helmut Schmidt University, Hamburg, Germany

A John Wiley & Sons, Ltd, Publication

Library of Congress Cataloging-in-Publication Data

Zölzer, Udo.
    Digital Audio Signal Processing / Udo Zölzer. 2nd ed.
        p. cm.
    Includes bibliographical reference and index.
    ISBN 978-0-470-99785-7 (cloth)
1. Sound–Recording and reproducing–Digital techniques. 2. Signal processing–Digital techniques. I. Title.
    TK7881.4.Z65 2008
    621.382'2–dc22   2008006095

A catalogue record for this book is available from the British Library.

ISBN 978-0-470-99785-7 (H/B)

Set in 10/12pt Times by Sunrise Setting Ltd, Torquay, England.

Contents

Preface to the Second Edition

This second edition represents a revised and extended version and offers an improved description besides new issues and extended references. The contents of this book are the basis of a lecture on *Digital Audio Signal Processing* at the Hamburg University of Technology (TU Hamburg-Harburg) and a lecture on *Multimedia Signal Processing* at the Helmut Schmidt University, Hamburg. For further studies you can find interactive audio demonstrations, exercises and Matlab examples on the web site

```
http://ant.hsu-hh.de/dasp/
```

Besides the basics of digital audio signal processing introduced in this second edition, further advanced algorithms for digital audio effects can be found in the book *DAFX – Digital Audio Effects* (Ed. U. Zölzer) with the related web site

```
http://www.dafx.de
```

My thanks go to Professor Dieter Leckschat, Dr. Gerald Schuller, Udo Ahlvers, Mijail Guillemard, Christian Helmrich, Martin Holters, Dr. Florian Keiler, Stephan Möller, Francois-Xavier Nsabimana, Christian Ruwwe, Harald Schorr, Dr. Oomke Weikert, Catja Wilkens and Christian Zimmermann.

Udo Zölzer
Hamburg, June 2008

Preface to the First Edition

Digital audio signal processing is employed in recording and storing music and speech signals, for sound mixing and production of digital programs, in digital transmission to broadcast receivers as well as in consumer products like CDs, DATs and PCs. In the latter case, the audio signal is in a digital form all the way from the microphone right up to the loudspeakers, enabling real-time processing with fast digital signal processors.

This book provides the basis of an advanced course in *Digital Audio Signal Processing* which I have been giving since 1992 at the Technical University Hamburg-Harburg. It is directed at students studying engineering, computer science and physics and also for professionals looking for solutions to problems in audio signal processing like in the fields of studio engineering, consumer electronics and multimedia. The mathematical and theoretical fundamentals of digital audio signal processing systems will be presented and typical applications with an emphasis on realization aspects will be discussed. Prior knowledge of systems theory, digital signal processing and multirate signal processing is taken as a prerequisite.

The book is divided into two parts. The first part (Chapters 1–4) presents a basis for hardware systems used in digital audio signal processing. The second part (Chapters 5–9) discusses algorithms for processing digital audio signals. Chapter 1 describes the course taken by an audio signal from its recording in a studio up to its reproduction at home. Chapter 2 contains a representation of signal quantization, dither techniques and spectral shaping of quantization errors used for reducing the nonlinear effects of quantization. In the end, a comparison is made between the fixed-point and floating-point number representations as well as their associated effects on format conversion and algorithms. Chapter 3 describes methods for AD/DA conversion of signals, starting with Nyquist sampling, methods for oversampling techniques and delta-sigma modulation. The chapter closes with a presentation of some circuit design of AD/DA converters. After an introduction to digital signal processors and digital audio interfaces, Chapter 4 describes simple hardware systems based on a single- and multiprocessor solutions. The algorithms introduced in the following Chapters 5–9 are, to a great extent, implemented in real-time on hardware platforms presented in Chapter 4. Chapter 5 describes digital audio equalizers. Apart from the implementation aspects of recursive audio filters, nonrecursive linear phase filters based on fast convolution and filter banks are introduced. Filter designs, parametric filter structures and precautions for reducing quantization errors in recursive filters are dealt with in detail. Chapter 6 deals with room simulation. Methods for simulation of artificial room impulse response and methods for approximation of measured impulse responses are discussed. In Chapter 7 the dynamic range control of audio signals is described. These methods are applied at several positions in the audio chain from the microphone up to the loudspeakers in order to adapt to the dynamics of the recording, transmission and listening environment. Chapter 8 contains a presentation of methods for synchronous and asynchronous sampling rate conversion. Efficient algorithms are described which are suitable for real-time processing as well as off-line processing. Both lossless and lossy audio coding are discussed in Chapter 9. Lossless audio coding is applied for storing of higher word-lengths. Lossy audio coding, on the other hand, plays a significant role in communication systems.

I would like to thank Prof. Fliege (University of Mannheim), Prof. Kammeyer (University of Bremen) and Prof. Heute (University of Kiel) for comments and support. I am also grateful to my colleagues at the TUHH and especially Dr. Alfred Mertins, Dr. Thomas Boltze, Dr. Bernd Redmer, Dr. Martin Schönle, Dr. Manfred Schusdziarra, Dr. Tanja Karp, Georg Dickmann, Werner Eckel, Thomas Scholz, Rüdiger Wolf, Jens Wohlers, Horst Zölzer, Bärbel Erdmann, Ursula Seifert and Dieter Gödecke. Apart from these, I would also like to say a word of gratitude to all those students who helped me in carrying out this work successfully.

Special thanks go to Saeed Khawaja for his help during translation and to Dr. Anthony Macgrath for proof-reading the text. I also would like to thank Jenny Smith, Colin McKerracher, Ian Stoneham and Christian Rauscher (Wiley).

My special thanks are directed to my wife Elke and my daughter Franziska.

Udo Zölzer
Hamburg, July 1997

Chapter 1

Introduction

It is hardly possible to make a start in the field of digital audio signal processing without having a first insight into the variety of technical devices and systems of audio technology. In this introductory chapter, the fields of application for digital audio signal processing are presented. Starting from recording in a studio or in a concert hall, the whole chain of signal processing is shown, up to the reproduction at home or in a car (see Fig. 1.1). The fields of application can be divided into the following areas:

- studio technology;

- digital transmission systems;

- storage media;

- audio components for home entertainment.

The basic principles of the above-mentioned fields of application will be presented as an overview in order to exhibit the uses of digital signal processing. Special technical devices and systems are outside the focus of this chapter. These devices and systems are strongly driven by the development of the computer technology with yearly changes and new devices based on new technologies. The goal of this introduction is a trend-independent presentation of the entire processing chain from the instrument or singer to the listener and consumer of music. The presentation of signal processing techniques and their algorithms will be discussed in the following chapters.

1.1   Studio Technology

While recording speech or music in a studio or in a concert hall, the analog signal from a microphone is first digitized, fed to a digital mixing console and then stored on a digital storage medium. A digital sound studio is shown in Fig. 1.2. Besides the analog sources (microphones), digital sources are fed to the digital mixing console over multichannel MADI interfaces [AES91]. Digital storage media like digital multitrack tape machines have been replaced by digital hard disc recording systems which are also connected via

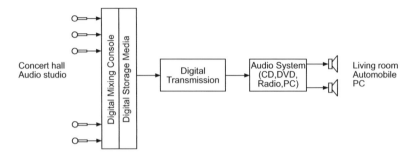

Figure 1.1 Signal processing for recording, storage, transmission and reproduction.

multichannel MADI interfaces to the mixing console. The final stereo mix is stored via a two-channel AES/EBU interface [AES92] on a two-channel MASTER machine. External appliances for effects or room simulators are also connected to the mixing console via a two-channel AES/EBU interface. All systems are synchronized by a MASTER clock reference. In digital audio technology, the sampling rates[1] $f_S = 48$ kHz for professional studio technology, $f_S = 44.1$ kHz for compact disc and $f_S = 32$ kHz for broadcasting applications are established. In addition, multiples of these sampling frequencies such as 88.2, 96, 176.4, and 192 kHz are used. The sound mixing console plays a central role in a digital sound studio. Figure 1.3 shows the functional units. The N input signals are processed individually. After level and panorama control, all signals are summed up to give a stereo mix. The summation is carried out several times so that other auxiliary stereo and/or mono signals are available for other purposes. In a sound channel (see Fig. 1.4), an equalizer unit (EQ), a dynamic unit (DYN), a delay unit (DEL), a gain element (GAIN) and a panorama element (PAN) are used. In addition to input and output signals in an audio channel, inserts as well as auxiliary or direct outputs are required.

1.2   Digital Transmission Systems

In this section digital transmission will be briefly explained. Besides the analog wireless broadcasting systems based on amplitude and frequency modulation, DAB[2] (Digital Audio Broadcasting) has been introduced in several countries [Hoe01]. On the other hand, the internet has pushed audio/video distribution, internet radio and video via cable networks.

Terrestrial Digital Broadcasting (DAB)

With the introduction of terrestrial digital broadcasting, the quality standards of a compact disc will be achieved for mobile and stationary reception of radio signals [Ple91]. Therefore, the data rate of a two-channel AES/EBU signal from a transmitting studio is reduced with the help of a source coder [Bra94] (see Fig. 1.5). Following the source coder (SC), additional information (AI) like the type of program (music/speech) and

---

[1]Data rate 16 bit × 48 kHz = 768 kbit/s; data rate (AES/EBU signal) 2 × (24 + 8) bit × 48 kHz = 3.072 Mbit/s; data rate (MADI signal) 56 × (24 + 8) bit × 48 kHz = 86.016 Mbit/s.

[2]http://www.worlddab.org/.

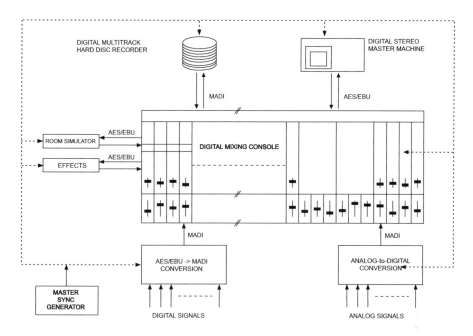

Figure 1.2 Digital sound studio.

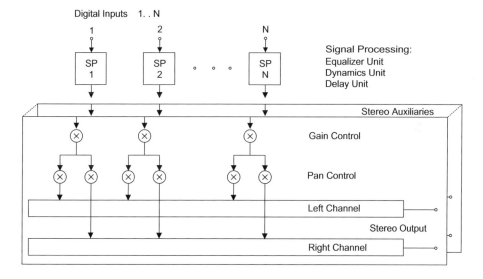

Figure 1.3 *N*-channel sound mixing console.

traffic information is added. A multicarrier technique is applied for digital transmission to stationary and mobile receivers. At the transmitter, several broadcasting programs are combined in a multiplexer (MUX) to form a multiplex signal. The channel coding and modulation is carried out by a multi-carrier transmission technique (*Coded Orthogonal Frequency Division Multiplex*, [Ala87, Kam92, Kam93, Tui93]).

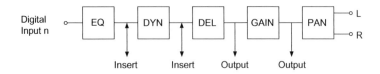

Figure 1.4 Sound channel.

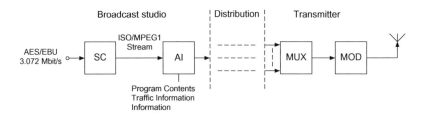

Figure 1.5 DAB transmitter.

The DAB receiver (Fig. 1.6) consists of the demodulator (DMOD), the demultiplexer (DMUX) and the source decoder (SD). The SD provides a linearly quantized PCM signal (*Pulse Code Modulation*). The PCM signal is fed over a *Digital-to-Analog Converter (DA Converter)* to an amplifier connected to loudspeakers.

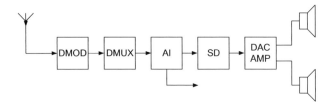

Figure 1.6 DAB receiver.

For a more detailed description of the DAB transmission technique, an illustration based on filter banks is presented (see Fig. 1.7). The audio signal at a data rate of 768 kbit/s is decomposed into subbands with the help of an analysis filter bank (AFB). Quantization and coding based on psychoacoustic models are carried out within each subband. The data reduction leads to a data rate of 96–192 kbit/s. The quantized subband signals are provided with additional information (header) and combined together in a frame. This so-called ISO-MPEG1 frame [ISO92] is first subjected to channel coding (CC). Time-interleaving (TIL) follows and will be described later on. The individual transmitting programs are combined in frequency multiplex (frequency-interleaving FIL) with a synthesis filter bank (SFB) into one broad-band transmitting signal. The synthesis filter bank has several complex-valued input signals and one complex-valued output signal. The real-valued band-pass signal is obtained by modulating with $e^{j\omega_c t}$ and taking the real part. At the receiver, the complex-valued base-band signal is obtained by demodulation followed by low-pass filtering. The complex-valued analysis filter bank provides the complex-valued band-pass signals from which the ISO-MPEG1 frame is formed after frequency and time deinterleaving and channel decoding. The PCM signal is combined using the synthesis filter bank after extracting the subband signals from the frame.

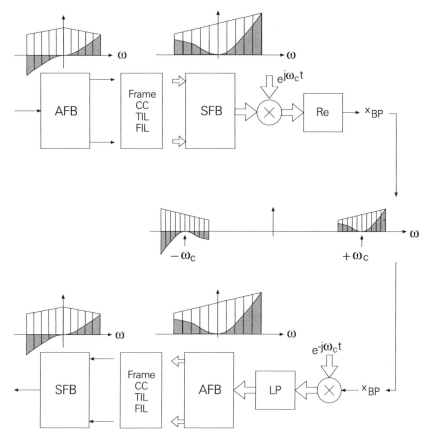

Figure 1.7 Filter banks within DAB.

DAB Transmission Technique. The special problems of mobile communications are dealt with using a combination of the OFDM transmission technique with DPSK modulation and time and frequency interleaving. Possible disturbances are minimized by consecutive channel coding. The schematic diagram in Fig. 1.8 shows the relevant subsystems.

For example, the transmission of a program P_1 which is delivered as an ISO-MPEG1 stream is shown in Fig. 1.8. The channel coding doubles the data rate. The typical characteristics of a mobile communication channel like time and frequency selectivity are handled by using time and frequency interleaving with the help of a multicarrier technique.

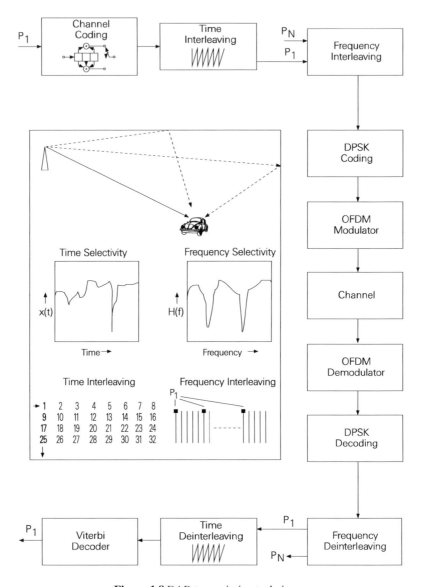

Figure 1.8 DAB transmission technique.

The burst disturbances of consecutive bits are reduced to single bit errors by spreading the bits over a longer period of time. The narrow-band disturbances affect only individual carriers by spreading the transmitter program P_1 in the frequency domain, i.e. distribution of transmitter programs of carrier frequencies at a certain displacement. The remaining disturbances of the mobile channel are suppressed with the help of channel coding, i.e. by adding redundancy, and decoding with a Viterbi decoder. The implementation of an OFDM transmission is discussed in the following.

OFDM Transmission. The OFDM transmission technique is shown in Fig. 1.9. The technique stands out owing to its simple implementation in the digital domain. The data sequence $c_t(k)$ which is to be transmitted, is written blockwise into a register of length $2M$. The complex numbers from $d_1(m)$ to $d_M(m)$ are formed from two consecutive bits (dibits). Here the first bit corresponds to the real part and the second to the imaginary part. The signal space shows the four states for the so-called QPSK [Kam92a, Pro89]. The vector $\mathbf{d}(m)$ is transformed with an inverse FFT (*Fast Fourier Transform*) into a vector $\mathbf{e}(m)$ which describes the values of the transmitted symbol in the time domain. The transmitted symbol $x_t(n)$ with period T_{sym} is formed by the transmission of the M complex numbers $e_i(m)$ at sampling period T_S. The real-valued band-pass signal is formed at high frequency after DA conversion of the quadrature signals, modulation by $e^{j\omega_c t}$ and by taking the real part. At the receiver, the transmitted symbol becomes a complex-valued sequence $x_r(n)$ by demodulation with $e^{-j\omega_c t}$ and AD conversion of the quadrature signal. M samples of the received sequence $x_r(n)$ are distributed over the M input values $f_i(m)$ and transformed into the frequency domain with the help of FFT. The resulting complex numbers $g_i(m)$ are again converted to dibits and provide the received sequence $c_r(k)$. Without the influence of the communication channel, the transmitted sequence can be reconstructed exactly.

OFDM Transmission with a Guard Interval. In order to describe the OFDM transmission with a guard interval, the schematic diagram in Fig. 1.10 is considered. The transmission of a symbol of length M over a channel with impulse response $h(n)$ of length L leads to a received signal $y(n)$ of length $M + L - 1$. This means that the received symbol is longer than the transmitted signal. The exact reconstruction of the transmitted symbol is disturbed because of the overlapping of received symbols. Reconstruction of the transmitted symbol is possible by cyclic continuation of the transmitted symbol. Here, the complex numbers from the vector $\mathbf{e}(m)$ are repeated so as to give a symbol period of $T_{\text{sym}} = (M + L)T_S$. Each of the transmitted symbols is, therefore, extended to a length of $M + L$. After transmission over a channel with impulse response of length L, the response of the channel is periodic with length M. After the initial transient state of the channel, i.e. after the L samples of the *guard interval*, the following M samples are written into a register. Since a time delay occurs between the start of the transmitted symbol and the sampling shifted by L displacements, it is necessary to shift the sequence of length M cyclically by L displacements. The complex values $g_i(m)$ do not correspond to the exact transmitted values $d_i(m)$ because of the transmission channel $h(n)$. However, there is no influence of neighboring carrier frequencies. Every received value $g_i(m)$ is weighted with the corresponding magnitude and phase of the channel at the specific carrier frequency. The influence of the communication channel can be eliminated by differential coding of consecutive dibits. The decoding process can be done according to $z_i(m) = g_i(m)g_i^*(m - 1)$. The dibit corresponds to the sign of the real and imaginary parts. The DAB transmission technique presented stands out owing to its simple implementation with the help of FFT algorithms. The extension of the transmitted symbol by a length L of the channel impulse response and the synchronization to collect the M samples out of the received symbol have still to be carried out. The length of the guard interval must be matched to the maximum echo delay of the multipath channel. Owing to differential coding of the transmitted sequence, an equalizer at the receiver is not necessary.

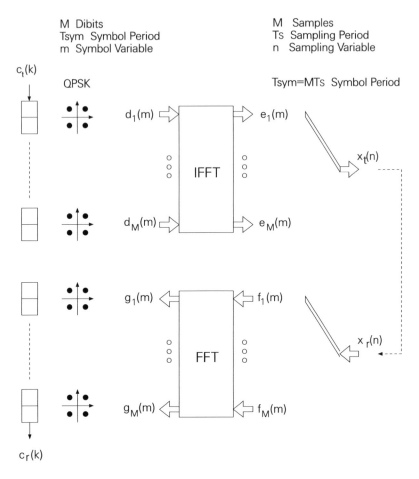

Figure 1.9 OFDM transmission.

Digital Radio Mondiale (DRM)

In the interest of national and international broadcasting stations a more widespread program delivery across regional or worldwide regions is of specific importance. This is accomplished by analog radio transmission in the frequency range below 30 MHz. The limited audio quality of the amplitude modulation technique (channel bandwidth 9–10 kHz) with an audio bandwidth of 4.5 kHz leads to a low acceptance rate for such kind of audio broadcasting. The introduction of the digital transmission technique Digital Radio Mondiale[3] will replace the existing analog transmission systems. The digital transmission is based on OFDM and the audio coding MPEG4-AAC in combination with SBR (Spectral Band Replication).

---

[3] http://www.drm.org.

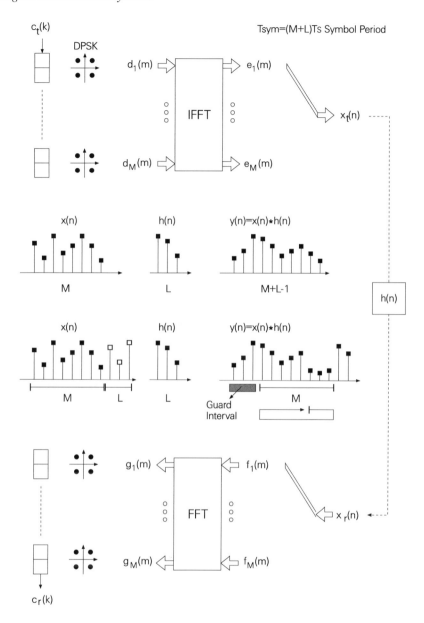

Figure 1.10 OFDM transmission with a guard interval.

Internet Audio

The growth of the internet offers new distribution possibilities for information, but especially for audio and video signals. The distribution of audio signals is mainly driven by the MP3 format (MPEG-1 Audio Layer III [Bra94]) or in proprietary formats of different companies. The compressed transmission is used because the data rate of home users is still low compared to lossless audio and video formats. Since the transmission is based on file transfer of packets, the data rates strongly depend on the providing server, the actual internet traffic and the access point of the home computer. A real-time transfer of high-quality music is still not possible. If the audio compression is high enough to achieve a just acceptable audio quality, a real-time transfer with a streaming technology is possible, since the file size is small and a transmission needs less time (see Fig. 1.11). For this a receiver needs a double memory filled with incoming packets of a coded audio file and a parallel running audio decoding. After decoding of a memory with a sufficiently long audio portion the memory is transferred to the sound card of the computer. During sound playback of the decoded audio signal further incoming packets are received and decoded. Packet loss can lead to interrupts in the audio signal. Several techniques for error concealment and protocols allow the transfer of coded audio.

Figure 1.11 Audio streaming via the internet.

1.3   Storage Media

Compact Disc

The technological advances in the semiconductor industry have led to economical storage media for digitally encoded information. Independently of developments in the computer industry, the compact disc system was introduced by Philips and Sony in 1982. The storage of digital audio data is carried out on an optical storage medium. The compact disc operates at a sampling rate of $f_S = 44.1$ kHz.[4] The essential specifications are summarized in Table 1.1.

R-DAT (*Rotary-head Digital Audio on Tape*)

The R-DAT system makes use of the heliscan method for two-channel recording. The available devices enable the recording of 16-bit PCM signals with all three sampling rates (Table 1.2) on a tape. R-DAT recorders are used in studio recording as well as in consumer applications.

MiniDisc and MP3 Format

Advanced coding techniques are based on psychoacoustics for data reduction. A widespread storage system is the MiniDisc by Sony. The Mini Disc system operates with the ATRAC technique (*Adaptive Transform Acoustic Coding*, [Tsu92]) and has a data rate of about

---

[4]$3 \times 490 \times 30$ Hz (NTSC) $= 3 \times 588 \times 25$ Hz (CCIR) $= 44.1$ kHz.

Table 1.1 Specifications of the CD system [Ben88].

Type of recording	
Signal recognition	Optical
Storage density	682 Mbit/in^2
Audio specification	
Number of channels	2
Duration	Approx. 60 min.
Frequency range	20–20 000 Hz
Dynamic range	>90 dB
THD	<0.01%
Signal format	
Sampling rate	44.1 kHz
Quantization	16-bit PCM (2's complement)
Pre-emphasis	None or 50/15 μs
Error Correction	CIRC
Data rate	2.034 Mbit/s
Modulation	EFM
Channel bit rate	4.3218 Mbit/s
Redundancy	30%
Mechanical specification	
Diameter	120 mm
Thickness	1.2 mm
Diameter of the inner hole	15 mm
Program range	50–116 mm
Reading speed	1.2–1.4 m/s
	500–200 r/min.

$2 \cdot 140$ kbit/s for a stereo channel. A magneto-optical storage medium is used for recording. The MP3 format was developed simultaneously, but the availability of recording and playback systems has taken a longer time. Simple MP3 recorders and playback systems are now available for the consumer market.

Super Audio Compact Disc (SACD)

The SACD was specified by Philips and Sony in 1999 as a further development of the compact disc with the objective of improved sound quality. The audio frequency range of 20 kHz is perceived as a limiting audio quality factor by some human beings, and the anti-aliasing and reconstruction filters may lead to *ringing* resulting from linear phase filters. This effect follows from short audio pulses leading to audible transients of the filters. In order to overcome these problems the audio bandwidth is extended to 100 kHz and the sampling frequency is increased to 2.8224 MHz (64 × 44.1 kHz). With this the filter specifications can be met with simple first-order filters. The quantization of the samples is based on a 1-bit quantizer within a delta-sigma converter structure which uses noise shaping

Table 1.2 Specifications of the R-DAT system [Ben88].

Type of recording	
Signal recognition	Magnetic
Storage capacity	2 GB
Audio specification	
Number of channels	2
Duration	Max. 120 min.
Frequency range	20–20 000 Hz
Dynamic range	>90 dB
THD	<0.01%
Signal format	
Sampling rate	48, 44.1, 32 kHz
Quantization	16-bit PCM (2's complement)
Error correction	CIRC
Channel coding	8/10 modulation
Data rate	2.46 Mbit/s
Channel bit rate	9.4 Mbit/s
Mechanical specification	
Tapewidth of magnet	3.8 mm
Thickness	13 μm
Diameter of head drum	3 cm
Revolutions per min.	2000 r/min.
Rel. track speed	3.133 m/s
	500–200 r/min.

(see Fig. 1.12). The 1-bit signal with 2.8224 MHz sampling frequency is denoted a DSD signal (Direct Stream Digital). The DA conversion of a DSD signal into an analog signal is accomplished with a simple analog first-order low-pass. The storage of DSD signals is achieved by a special compact disc (Fig. 1.13) with a CD layer in PCM format and an HD layer (High Density) with a DVD 4.38 GByte layer. The HD layer stores a stereo signal in 1-bit DSD format and a 6-channel 1-bit signal with a lossless compression technique (Direct Stream Transfer DST) [Jan03]. The CD layer of the SACD can be replayed with a conventional CD player, whereas special SACD players can replay the HD layer. An extensive discussion of 1-bit delta-sigma techniques can be found in [Lip01a, Lip01b, Van01, Lip02, Van04].

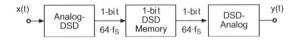

Figure 1.12 SACD system.

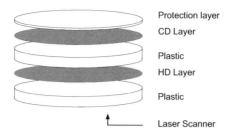

Figure 1.13 Layer of the SACD.

Digital Versatile Disc – Audio (DVD-A)

To increase the storage capacity of the CD the Digital Versatile Disc (DVD) was developed. The physical dimensions are identical to the CD. The DVD has two layers with one or two sides, and the storage capacity per side has been increased. For a one-sided version for audio applications the storage capacity is 4.7 GB. A comparison of specifications for different disc media is shown in Table 1.3. Besides stereo signals with different sampling rates and word-lengths a variety of multi-channel formats can be stored. For data reduction a lossless compression technique, MLP (Meridian Lossless Packing), is applied. The improved audio quality compared to the CD audio is based on the higher sampling rates and word-lengths and the multichannel features of the DVD-A.

Table 1.3 Specifications of CD, SACD and DVD-A.

Parameter	CD	SACD	DVD-A
Coding	16-bit PCM	1-bit DSD	16-/20-/24-bit PCM
Sampling rate	44.1 kHz	2.8224 MHz	44.1/48/88.2/96/176.4/192 kHz
Channels	2	2–6	1–6
Compression	No	Yes (DST)	Yes (MLP)
Recording time	74 min.	70–80 min.	62–843 min.
Frequency range	20–20 000 Hz	20–100 000 Hz	20–96 000 Hz
Dynamic range	96 dB	120 dB	144 dB
Copy protection	No	Yes	Yes

1.4 Audio Components at Home

Domestic digital storage media are already in use, like compact discs, personal computers and MP3 players, which have digital outputs, and can be connected to digital post-processing systems right up to the loudspeakers. The individual tone control consists of the following processing.

Equalizer

Spectral modification of the music signal in amplitude and phase and the automatic correction of the frequency response from loudspeaker to listening environment are desired.

Room Simulation

The simulation of room impulse responses and the processing of music signals with special room impulse response are used to give an impression of a room like a concert hall, a cathedral or a jazz club.

Surround Systems

Besides the reproduction of stereo signals from a CD over two frontal loudspeakers, more than two channels will be recorded in the prospective digital recording systems [Lin93]. This is already illustrated in the sound production for cinema movies where, besides the stereo signal (L, R), a middle channel (M) and two additional room signals (L_B, R_B) are recorded. These *surround* systems are also used in the prospective digital television systems. The *ambisonics* technique [Ger85] is a recording technique that allows three-dimensional recording and reproduction of sound.

Digital Amplifier Concepts

The basis of a digital amplifier is pulse width modulation as shown in Fig. 1.14. With the help of a fast counter, a pulse width modulated signal is formed out of the w-bit linearly quantized signal. Single-sided and double-sided modulated conversion are used and they are represented by two and three states, respectively. Single-sided modulation (2 states, -1 and $+1$) is performed by a counter which counts upward from zero with multiples of the sampling rate. The number range of the PCM signal from -1 to $+1$ is directly mapped onto the counter. The duration of the pulse width is controlled by a comparator. For pulse width modulation with three states $(-1, 0, +1)$, the sign of the PCM signal determines the state. The pulse width is determined by a mapping of the number range from 0 to 1 onto a counter. For double-sided modulation, an upward/downward counter is needed which has to be clocked at twice the rate compared with single-sided modulation. The allocation of pulse widths is shown in Fig. 1.14. In order to reduce the clock rate for the counter, pulse width modulation is carried out after oversampling (*Oversampling*) and noise shaping (*Noise Shaping*) of the quantization error (see Fig. 1.15, [Gol90]). Thus the clock rate of the counter is reduced to 180.6 MHz. The input signal is first upsampled by a factor of 16 and then quantized to 8-bits with third-order noise shaping. The use of pulse shaping with delta-sigma modulation is shown in Fig. 1.16 [And92]. Here a direct conversion of the delta-sigma modulated 1-bit signal is performed. The pulse converter shapes the envelope of the serial data bits. The low-pass filter reconstructs the analog signal. In order to reduce nonlinear distortion, the output signal is fed back (see Fig. 1.17, [Klu92]). New methods for the generation of pulse width modulation try to reduce the clock rates and the high frequency components [Str99, Str01].

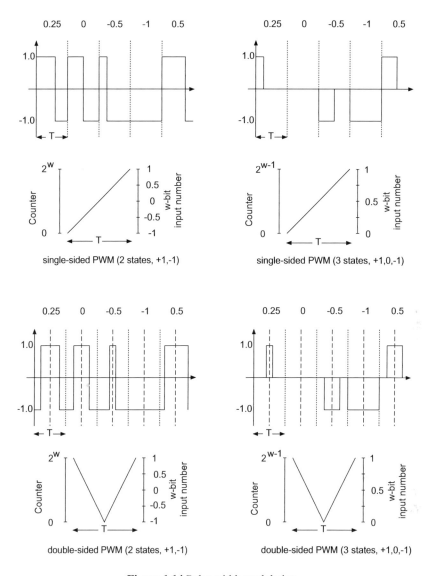

Figure 1.14 Pulse width modulation.

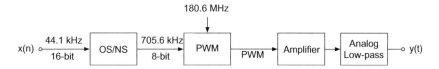

Figure 1.15 Pulse width modulation with oversampling and noise shaping.

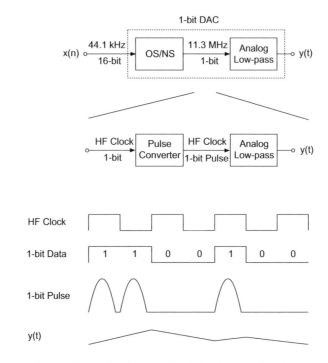

Figure 1.16 Pulse shaping after delta-sigma modulation.

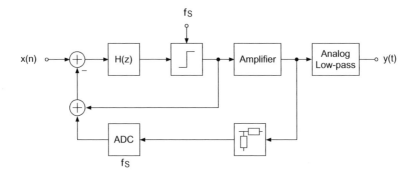

Figure 1.17 Delta-sigma modulated amplifier with feedback.

Digital Crossover

In order to perform digital crossovers for loudspeakers, a linear phase decomposition of the signal with a special filter bank [Zöl92] is done (Fig. 1.18). In a first step, the input signal is decomposed into its high-pass and low-pass components and the high-pass signal is fed to a DAC over a delay unit. In the next step, the low-pass signal is further decomposed. The individual band-pass signals and the low-pass signal are then fed to the respective loudspeaker. Further developments for the control of loudspeakers can be found in [Kli94, Kli98a, Kli98b, Mül99].

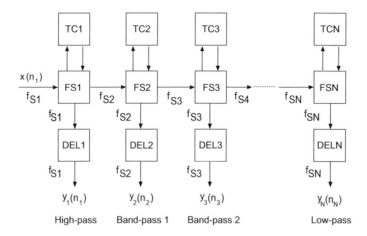

Figure 1.18 Digital crossover (FS$_i$ frequency splitting, TC$_i$ transition bandwidth control, DEL$_i$ delay).

References

[AES91]   AES10-1991 (ANSI S4.43-1991): AES Recommended Practice for Digital Audio Engineering – Serial Multichannel Audio Digital Interface (MADI).

[AES92]   AES3-1992 (ANSI S4.40-1992): AES Recommended Practice for Digital Audio Engineering – Serial Transmission Format for Two-Channel Linearly Represented Digital Audio.

[Ala87]   M. Alard, R. Lasalle: *Principles of Modulation and Channel Coding for Digital Broadcasting for Mobile Receivers*, EBU Review, No. 224, pp. 168–190, August 1987.

[And92]   M. Andersen: *New Principle for Digital Audio Power Amplifiers*, Proc. 92nd AES Convention, Preprint No. 3226, Vienna, 1992.

[Ben88]   K. B. Benson: *Audio Engineering Handbook*, McGraw-Hill, New York, 1988.

[Bra94]   K. Brandenburg, G. Stoll: *ISO/MPEG-1 Audio: A Generic Standard for Coding of High Quality Digital Audio*, J. Audio Eng. Soc., Vol. 42, No. 10, pp. 780–792, October 1994.

[Ger85]   M. A. Gerzon: *Ambisonics in Multichannel Broadcasting and Video*, J. Audio Eng. Soc., Vol. 33, No. 11, pp. 859–871, November 1985.

[Gol90]   J. M. Goldberg, M. B. Sandler: *New Results in PWM for Digital Power Amplification*, Proc. 89th AES Convention, Preprint No. 2959, Los Angeles, 1990.

[Hoe01]   W. Hoeg, T. Lauterbach: *Digital Audio Broadcasting*, John Wiley & Sons Ltd, Chichester, 2001.

[ISO92]   ISO/IEC 11172-3: *Coding of Moving Pictures and Associated Audio for Digital Storage Media at up to 1.5 Mbits/s – Audio Part*, International Standard, 1992.

[Jan03]   E. Janssen, D. Reefman: *Super Audio CD: An Introduction*, IEEE Signal Processing Magazine, pp. 83–90, July 2003.

[Kam92]   K. D. Kammeyer, U. Tuisel, H. Schulze, H. Bochmann: *Digital Multicarrier-Transmission of Audio Signals over Mobile Radio Channels*, Europ. Trans. on Telecommun. ETT, Vol. 3, pp. 243–254, May–June 1992.

[Kam93]   K. D. Kammeyer, U. Tuisel: *Synchronisationsprobleme in digitalen Multi-trägersystemen*, Frequenz, Vol. 47, pp. 159–166, May 1993.

[Kli94]   W. Klippel: *Das nichtlineare Übertragungsverhalten elektroakustischer Wandler*, Habilitationsschrift, Technische Universität Dresden, 1994.

[Kli98a]   W. Klippel: *Direct Feedback Linearization of Nonlinear Loudspeaker Systems*, J. Audio Eng. Soc., Vol. 46, No. 6, pp. 499–507, 1998.

[Kli98b]   W. Klippel: *Adaptive Nonlinear Control of Loudspeaker Systems*, J. Audio Eng. Soc., Vol. 46, No. 11, pp. 939–954, 1998.

[Klu92]   J. Klugbauer-Heilmeier: *A Sigma Delta Modulated Switching Power Amp*, Proc. 92nd AES Convention, Preprint No. 3227, Vienna, 1992.

[Lec92]   D. Leckschat: *Verbesserung der Wiedergabequalität von Lautsprechern mit Hilfe von Digitalfiltern*, Dissertation, RWTH Aachen, 1992.

[Lin93]   B. Link, D. Mandell: *A DSP Implementation of a Pro Logic Surround Decoder*, Proc. 95th AES Convention, Preprint No. 3758, New York, 1993.

[Lip01a]   S. P. Lipshitz, J. Vanderkooy: *Why 1-Bit Sigma-Delta Conversion is Unsuitable for High-Quality Applications*, Proc. 110th Convention of the Audio Engineering Society, Preprint No. 5395, Amsterdam, 2001.

[Lip01b]   S. P. Lipshitz. J. Vanderkooy: *Towards a Better Understanding of 1-Bit Sigma-Delta Modulators – Part 2*, Proc. 111th Convention of the Audio Engineering Society, Preprint No. 5477, New York, 2001

[Lip02]   S. P. Lipshitz, J. Vanderkooy: *Towards a Better Understanding of 1-Bit Sigma-Delta Modulators – Part 3*, Proc. 112th Convention of the Audio Engineering Society, Preprint No. 5620, Munich, 2002.

[Mül99]   S. Müller: *Digitale Signalverarbeitung für Lautsprecher*, Dissertation, RWTH Aachen, 1999.

[Ple91]   G. Plenge: *DAB – Ein neues Hörrundfunksystem – Stand der Entwicklung und Wege zu seiner Einführung*, Rundfunktechnische Mitteilungen, Vol. 35, No. 2, pp. 46–66, 1991.

[Str99]   M. Streitenberger, H. Bresch, W. Mathis: *A New Concept for High Performance Class-D Audio Amplification*, Proc. AES 106th Convention, Preprint No. 4941, Munich, 1999.

[Str01]   M. Streitenberger, F. Felgenhauer, H. Bresch, W. Mathis: *Zero Position Coding (ZePoC) – A Generalised Concept of Pulse-Length Modulated Signals and its Application to Class-D Audio Power Amplifiers*, Proc. AES 110th Convention, Preprint No. 5365, Amsterdam, 2001.

[Tsu92]   K. Tsutsui, H. Suzuki, O. Shimoyoshi, M. Sonohara, K. Akagiri, R. Heddle: *ATRAC: Adaptive Transform Coding for MiniDisc*, Proc. 91st AES Convention, Preprint No. 3216, New York, 1991.

[Tui93]   U. Tuisel: *Multiträgerkonzepte für die digitale, terrestrische Hörrundfunkübertragung*, Dissertation, TU Hamburg-Harburg, 1993.

[Van01]   J. Vanderkooy, S. P. Lipshitz: *Towards a Better Understanding of 1-Bit Sigma-Delta Modulators – Part 1*, Proc. 110th Convention of the Audio Engineering Society, Preprint No. 5398, Amsterdam, 2001.

[Van04]   J. Vanderkooy, S. P. Lipshitz: *Towards a Better Understanding of 1-Bit Sigma-Delta Modulators – Part 4*, Proc. 116th Convention of the Audio Engineering Society, Preprint No. 6093, Berlin, 2004.

[Zöl92]   U. Zölzer, N. Fliege: *Logarithmic Spaced Analysis Filter Bank for Multiple Loudspeaker Channels*, Proc. 93rd AES Convention, Preprint No. 3453, San Francisco, 1992.

Chapter 2

Quantization

Basic operations for AD conversion of a continuous-time signal $x(t)$ are the sampling and quantization of $x(n)$ yielding the quantized sequence $x_Q(n)$ (see Fig. 2.1). Before discussing AD/DA conversion techniques and the choice of the sampling frequency $f_S = 1/T_S$ in Chapter 3 we will introduce the quantization of the samples $x(n)$ with finite number of bits. The digitization of a sampled signal with continuous amplitude is called quantization. The effects of quantization starting with the classical quantization model are discussed in Section 2.1. In Section 2.2 dither techniques are presented which, for low-level signals, linearize the process of quantization. In Section 2.3 spectral shaping of quantization errors is described. Section 2.4 deals with number representation for digital audio signals and their effects on algorithms.

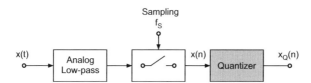

Figure 2.1 AD conversion and quantization.

2.1 Signal Quantization

2.1.1 Classical Quantization Model

Quantization is described by *Widrow's quantization theorem* [Wid61]. This says that a quantizer can be modeled (see Fig. 2.2) as the addition of a uniform distributed random signal $e(n)$ to the original signal $x(n)$ (see Fig. 2.2, [Wid61]). This additive model,

$$x_Q(n) = x(n) + e(n), \tag{2.1}$$

is based on the difference between quantized output and input according to the error signal

$$e(n) = x_Q(n) - x(n). \tag{2.2}$$

Digital Audio Signal Processing   Second Edition   Udo Zölzer
© 2008 John Wiley & Sons, Ltd

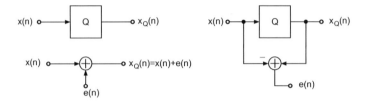

Figure 2.2 Quantization.

This linear model of the output $x_Q(n)$ is only then valid when the input amplitude has a wide dynamic range and the quantization error $e(n)$ is not correlated with the signal $x(n)$. Owing to the statistical independence of consecutive quantization errors the autocorrelation of the error signal is given by $r_{EE}(m) = \sigma_E^2 \cdot \delta(m)$, yielding a power density spectrum $S_{EE}(e^{j\Omega}) = \sigma_E^2$.

The nonlinear process of quantization is described by a nonlinear characteristic curve as shown in Fig. 2.3a, where Q denotes the quantization step. The difference between output and input of the quantizer provides the quantization error $e(n) = x_Q(n) - x(n)$, which is shown in Fig. 2.3b. The uniform probability density function (PDF) of the quantization error is given (see Fig. 2.3b) by

$$p_E(e) = \frac{1}{Q}\mathrm{rect}\left(\frac{e}{Q}\right). \tag{2.3}$$

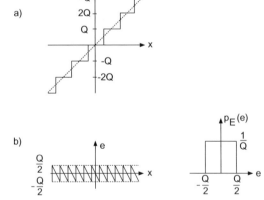

Figure 2.3 (a) Nonlinear characteristic curve of a quantizer. (b) Quantization error e and its probability density function (PDF) $p_E(e)$.

The mth moment of a random variable E with a PDF $p_E(e)$ is defined as the expected value of E^m:

$$E\{E^m\} = \int_{-\infty}^{\infty} e^m p_E(e)\,de. \tag{2.4}$$

For a uniform distributed random process, as in (2.3), the first two moments are given by

$$m_E = E\{E\} = 0 \quad \text{mean value} \tag{2.5}$$

$$\sigma_E^2 = E\{E^2\} = \frac{Q^2}{12} \quad \text{variance.} \tag{2.6}$$

The signal-to-noise ratio

$$\text{SNR} = 10 \log_{10}\left(\frac{\sigma_X^2}{\sigma_E^2}\right) \text{ dB} \tag{2.7}$$

is defined as the ratio of signal power σ_X^2 to error power σ_E^2.

For a quantizer with input range $\pm x_{\max}$ and word-length w, the quantization step size can be expressed as

$$Q = 2x_{\max}/2^w. \tag{2.8}$$

By defining a peak factor,

$$P_F = \frac{x_{\max}}{\sigma_X} = \frac{2^{w-1} Q}{\sigma_X}, \tag{2.9}$$

the variances of the input and the quantization error can be written as

$$\sigma_X^2 = \frac{x_{\max}^2}{P_F^2}, \tag{2.10}$$

$$\sigma_E^2 = \frac{Q^2}{12} = \frac{1}{12} \frac{x_{\max}^2}{2^{2w}} 2^2 = \frac{1}{3} x_{\max}^2 2^{-2w}. \tag{2.11}$$

The signal-to-noise ratio is then given by

$$\text{SNR} = 10 \log_{10}\left(\frac{x_{\max}^2/P_F^2}{\frac{1}{3} x_{\max}^2 2^{-2w}}\right) = 10 \log_{10}\left(2^{2w} \frac{3}{P_F^2}\right)$$

$$= 6.02w - 10 \log_{10}(P_F^2/3) \text{ dB.} \tag{2.12}$$

A sinusoidal signal (PDF as in Fig. 2.4) with $P_F = \sqrt{2}$ gives

$$\text{SNR} = 6.02w + 1.76 \text{ dB.} \tag{2.13}$$

For a signal with uniform PDF (see Fig. 2.4) and $P_F = \sqrt{3}$ we can write

$$\text{SNR} = 6.02w \text{ dB,} \tag{2.14}$$

and for a Gaussian distributed signal (probability of overload $<10^{-5}$ leads to $P_F = 4.61$, see Fig. 2.5), it follows that

$$\text{SNR} = 6.02w - 8.5 \text{ dB.} \tag{2.15}$$

It is obvious that the signal-to-noise ratio depends on the PDF of the input. For digital audio signals that exhibit nearly Gaussian distribution, the maximum signal-to-noise ratio for given word-length w is 8.5 dB lower than the *rule-of-thumb* formula (2.14) for the signal-to-noise ratio.

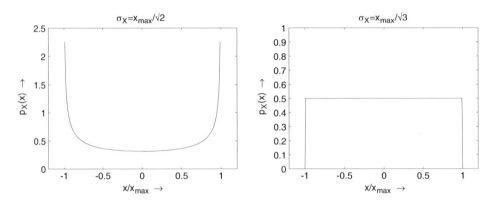

Figure 2.4 Probability density function (sinusoidal signal and signal with uniform PDF).

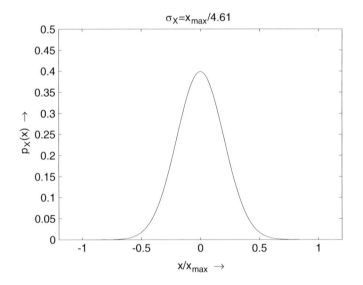

Figure 2.5 Probability density function (signal with Gaussian PDF).

2.1.2   Quantization Theorem

The statement of the *quantization theorem* for amplitude sampling (digitizing the amplitude) of signals was given by Widrow [Wid61]. The analogy for digitizing the time axis is the *sampling theorem* due to Shannon [Sha48]. Figure 2.6 shows the amplitude quantization and the time quantization. First of all, the PDF of the output signal of a quantizer is determined in terms of the PDF of the input signal. Both probability density functions are shown in Fig. 2.7. The respective characteristic functions (Fourier transform of a PDF) of the input and output signals form the basis for Widrow's quantization theorem.

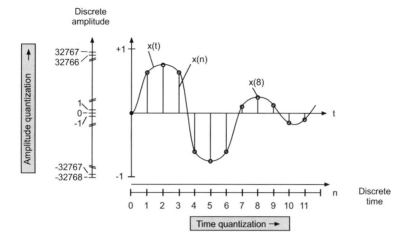

Figure 2.6 Amplitude and time quantization.

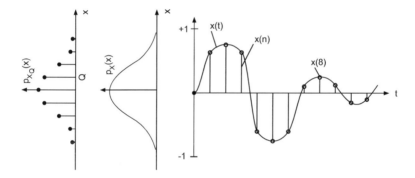

Figure 2.7 Probability density function of signal $x(n)$ and quantized signal $x_Q(n)$.

First-order Statistics of the Quantizer Output

Quantization of a continuous-amplitude signal x with PDF $p_X(x)$ leads to a discrete-amplitude signal y with PDF $p_Y(y)$ (see Fig. 2.8). The continuous PDF of the input is sampled by integrating over all quantization intervals (zone sampling). This leads to a discrete PDF of the output.

In the quantization intervals, the discrete PDF of the output is determined by the probability

$$W[kQ] = W\left[-\frac{Q}{2} + kQ \le x < \frac{Q}{2} + kQ\right] = \int_{-Q/2+kQ}^{Q/2+kQ} p_X(x)\,dx. \qquad (2.16)$$

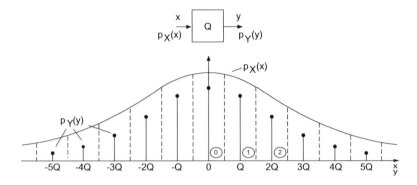

Figure 2.8 Zone sampling of the PDF.

For the intervals $k = 0, 1, 2$, it follows that

$$p_Y(y) = \delta(0) \int_{-Q/2}^{Q/2} p_X(x)\, dx, \qquad -\frac{Q}{2} \leq y < \frac{Q}{2},$$

$$= \delta(y - Q) \int_{-Q/2+Q}^{Q/2+Q} p_X(x)\, dx, \qquad -\frac{Q}{2} + Q \leq y < \frac{Q}{2} + Q,$$

$$= \delta(y - 2Q) \int_{-Q/2+2Q}^{Q/2+2Q} p_X(x)\, dx, \qquad -\frac{Q}{2} + 2Q \leq y < \frac{Q}{2} + 2Q.$$

The summation over all intervals gives the PDF of the output

$$p_Y(y) = \sum_{k=-\infty}^{\infty} \delta(y - kQ) W(kQ) \tag{2.17}$$

$$= \sum_{k=-\infty}^{\infty} \delta(y - kQ) W(y), \tag{2.18}$$

where

$$W(kQ) = \int_{-Q/2+kQ}^{Q/2+kQ} p_X(x)\, dx, \tag{2.19}$$

$$W(y) = \int_{-\infty}^{\infty} \text{rect}\left(\frac{y - x}{Q}\right) p_X(x)\, dx \tag{2.20}$$

$$= \text{rect}\left(\frac{y}{Q}\right) * p_X(y). \tag{2.21}$$

Using

$$\delta_Q(y) = \sum_{k=-\infty}^{\infty} \delta(y - kQ), \tag{2.22}$$

the PDF of the output is given by

$$p_Y(y) = \delta_Q(y) \left[\text{rect}\left(\frac{y}{Q}\right) \star p_X(y) \right]. \tag{2.23}$$

The PDF of the output can hence be determined by convolution of a rect function [Lip92] with the PDF of the input. This is followed by an *amplitude sampling* with resolution Q as described in (2.23) (see Fig. 2.9).

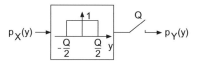

Figure 2.9 Determining PDF of the output.

Using $FT\{f_1(t) \cdot f_2(t)\} = \frac{1}{2\pi} F_1(j\omega) * F_2(j\omega)$, the characteristic function (Fourier transform of $p_Y(y)$) can be written, with $u_o = 2\pi/Q$, as

$$P_Y(ju) = \frac{1}{2\pi} u_o \sum_{k=-\infty}^{\infty} \delta(u - ku_o) * \left[Q \frac{\sin\left(u\frac{Q}{2}\right)}{u\frac{Q}{2}} \cdot P_X(ju) \right] \qquad (2.24)$$

$$= \sum_{k=-\infty}^{\infty} \delta(u - ku_o) * \left[\frac{\sin\left(u\frac{Q}{2}\right)}{u\frac{Q}{2}} \cdot P_X(ju) \right] \qquad (2.25)$$

$$\boxed{P_Y(ju) = \sum_{k=-\infty}^{\infty} P_X(ju - jku_o) \frac{\sin\left[(u - ku_o)\frac{Q}{2}\right]}{(u - ku_o)\frac{Q}{2}}.} \qquad (2.26)$$

Equation (2.26) describes the sampling of the continuous PDF of the input. If the quantization frequency $u_o = 2\pi/Q$ is twice the highest frequency of the characteristic function $P_X(ju)$ then periodically recurring spectra do not overlap. Hence, a reconstruction of the PDF of the input $p_X(x)$ from the quantized PDF of the output $p_Y(y)$ is possible (see Fig. 2.10). This is known as Widrow's *quantization theorem*. Contrary to the first sampling theorem (Shannon's sampling theorem, ideal amplitude sampling in the time domain) $F^A(j\omega) = \frac{1}{T} \sum_{k=-\infty}^{\infty} F(j\omega - jk\omega_o)$, it can be observed that there is an additional multiplication of the periodically characteristic function with $\frac{\sin\left[(u-ku_o)\frac{Q}{2}\right]}{(u-ku_o)\frac{Q}{2}}$ (see (2.26)).

If the base-band of the characteristic function ($k = 0$)

$$P_Y(ju) = P_X(ju) \underbrace{\frac{\sin\left(u\frac{Q}{2}\right)}{u\frac{Q}{2}}}_{P_E(ju)} \qquad (2.27)$$

is considered, it is observed that it is a product of two characteristic functions. The multiplication of characteristic functions leads to the convolution of PDFs from which the addition of two statistically independent signals can be concluded. The characteristic function of the quantization error is thus

$$P_E(ju) = \frac{\sin\left(u\frac{Q}{2}\right)}{u\frac{Q}{2}}, \qquad (2.28)$$

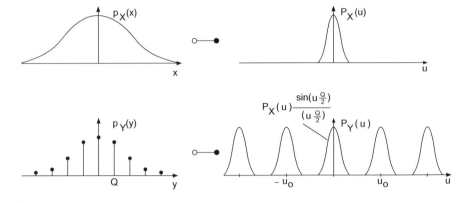

Figure 2.10 Spectral representation.

and the PDF

$$p_E(e) = \frac{1}{Q}\text{rect}\left(\frac{e}{Q}\right) \tag{2.29}$$

(see Fig. 2.11).

Figure 2.11 PDF and characteristic function of quantization error.

The modeling of the quantization process as an addition of a statistically independent noise signal to the input signal leads to a continuous PDF of the output (see Fig. 2.12, convolution of PDFs and sampling in the interval Q gives the discrete PDF of the output). The PDF of the discrete-valued output comprises Dirac pulses at distance Q with values equal to the continuous PDF (see (2.23)). Only if the quantization theorem is valid can the continuous PDF be reconstructed from the discrete PDF.

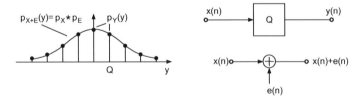

Figure 2.12 PDF of model.

In many cases, it is not necessary to reconstruct the PDF of the input. It is sufficient to calculate the moments of the input from the output. The mth moment can be expressed in terms of the PDF or the characteristic function:

$$E\{Y^m\} = \int_{-\infty}^{\infty} y^m \, p_Y(y) \, dy \tag{2.30}$$

$$= (-j)^m \frac{d^m \, P_Y(ju)}{du^m}\bigg|_{u=0}. \tag{2.31}$$

If the quantization theorem is satisfied then the periodic terms in (2.26) do not overlap and the mth derivative of $P_Y(ju)$ is solely determined by the base-band[1] so that, with (2.26),

$$E\{Y^m\} = (-j)^m \frac{d^m}{du^m} P_X(ju) \frac{\sin\left(u\frac{Q}{2}\right)}{u\frac{Q}{2}}\bigg|_{u=0}. \tag{2.32}$$

With (2.32), the first two moments can be determined as

$$m_Y = E\{Y\} = E\{X\}, \tag{2.33}$$

$$\sigma_Y^2 = E\{Y^2\} = \underbrace{E\{X^2\}}_{\sigma_X^2} + \underbrace{\frac{Q^2}{12}}_{\sigma_E^2}. \tag{2.34}$$

Second-order Statistics of Quantizer Output

In order to describe the properties of the output in the frequency domain, two output values Y_1 (at time n_1) and Y_2 (at time n_2) are considered [Lip92]. For the joint density function,

$$p_{Y_1 Y_2}(y_1, y_2) = \delta_{QQ}(y_1, y_2) \left[\text{rect}\left(\frac{y_1}{Q}, \frac{y_2}{Q}\right) \star p_{X_1 X_2}(y_1, y_2) \right], \tag{2.35}$$

with

$$\delta_{QQ}(y_1, y_2) = \delta_Q(y_1) \cdot \delta_Q(y_2) \tag{2.36}$$

and

$$\text{rect}\left(\frac{y_1}{Q}, \frac{y_2}{Q}\right) = \text{rect}\left(\frac{y_1}{Q}\right) \cdot \text{rect}\left(\frac{y_2}{Q}\right). \tag{2.37}$$

For the two-dimensional Fourier transform, it follows that

$$P_{Y_1 Y_2}(ju_1, ju_2) = \sum_{k=-\infty}^{\infty} \sum_{l=-\infty}^{\infty} \delta(u_1 - ku_o)\delta(u_2 - lu_o)$$

$$\star \left[\frac{\sin\left(u_1\frac{Q}{2}\right)}{u_1\frac{Q}{2}} \cdot \frac{\sin\left(u_2\frac{Q}{2}\right)}{u_2\frac{Q}{2}} \cdot P_{X_1 X_2}(ju_1, ju_2) \right] \tag{2.38}$$

$$= \sum_{k=-\infty}^{\infty} \sum_{l=-\infty}^{\infty} P_{X_1 X_2}(ju_1 - jku_o, ju_2 - jlu_o)$$

$$\cdot \frac{\sin\left[(u_1 - ku_o)\frac{Q}{2}\right]}{(u_1 - ku_o)\frac{Q}{2}} \cdot \frac{\sin\left[(u_2 - lu_o)\frac{Q}{2}\right]}{(u_2 - lu_o)\frac{Q}{2}}. \tag{2.39}$$

---

[1]This is also valid owing to the weaker condition of Sripad and Snyder [Sri77] discussed in the next section.

Similar to the one-dimensional quantization theorem, a two-dimensional theorem [Wid61] can be formulated: the joint density function of the input can be reconstructed from the joint density function of the output, if $P_{X_1 X_2}(ju_1, ju_2) = 0$ for $u_1 \geq u_o/2$ and $u_2 \geq u_o/2$. Here again, the moments of the joint density function can be calculated as follows:

$$E\{Y_1^m Y_2^n\} = (-j)^{m+n} \frac{\partial^{m+n}}{\partial u_1^m \partial u_2^n} P_{X_1 X_2}(ju_1, ju_2) \frac{\sin\left(u_1 \frac{Q}{2}\right)}{u_1 \frac{Q}{2}} \frac{\sin\left(u_2 \frac{Q}{2}\right)}{u_2 \frac{Q}{2}} \Bigg|_{u_1=0, u_2=0}. \tag{2.40}$$

From this, the autocorrelation function with $m = n_2 - n_1$ can be written as

$$r_{YY}(m) = E\{Y_1 Y_2\}(m) = \begin{cases} E\{X^2\} + \dfrac{Q^2}{12}, & m = 0, \\ E\{X_1 X_2\}(m), & \text{elsewhere} \end{cases} \tag{2.41}$$

(for $m = 0$ we obtain (2.34)).

2.1.3   Statistics of Quantization Error

First-order Statistics of Quantization Error

The PDF of the quantization error depends on the PDF of the input and is dealt with in the following. The quantization error $e = x_Q - x$ is restricted to the interval $\left[-\frac{Q}{2}, \frac{Q}{2}\right]$. It depends linearly on the input (see Fig. 2.13). If the input value lies in the interval $\left[-\frac{Q}{2}, \frac{Q}{2}\right]$ then the error is $e = 0 - x$. For the PDF we obtain $p_E(e) = p_X(e)$. If the input value lies in the interval $\left[-\frac{Q}{2} + Q, \frac{Q}{2} + Q\right]$ then the quantization error is $e = Q\lfloor Q^{-1}x + 0.5\rfloor - x$ and is again restricted to $\left[-\frac{Q}{2}, \frac{Q}{2}\right]$. The PDF of the quantization error is consequently $p_E(e) = p_X(e + Q)$ and is added to the first term. For the sum over all intervals we can write

$$p_E(e) = \begin{cases} \displaystyle\sum_{k=-\infty}^{\infty} p_X(e - kQ), & -\dfrac{Q}{2} \leq e < \dfrac{Q}{2}, \\ 0, & \text{elsewhere.} \end{cases} \tag{2.42}$$

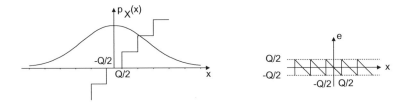

Figure 2.13 Probability density function and quantization error.

Because of the restricted values of the variable of the PDF, we can write

$$p_E(e) = \text{rect}\left(\frac{e}{Q}\right) \sum_{k=-\infty}^{\infty} p_X(e - kQ) \tag{2.43}$$

$$= \text{rect}\left(\frac{e}{Q}\right)[p_X(e) * \delta_Q(e)]. \tag{2.44}$$

The PDF of the quantization error is determined by the PDF of the input and can be computed by shifting and windowing a zone. All individual zones are summed up for calculating the PDF of the quantization error [Lip92]. A simple graphical interpretation of this overlapping is shown in Fig. 2.14. The overlapping leads to a uniform distribution of the quantization error if the input PDF $p_X(x)$ is spread over a sufficient number of quantization intervals.

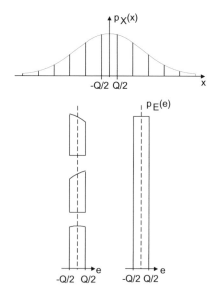

Figure 2.14 Probability density function of the quantization error.

For the Fourier transform of the PDF from (2.44) it follows that

$$P_E(ju) = \frac{1}{2\pi} Q \frac{\sin\left(u\frac{Q}{2}\right)}{u\frac{Q}{2}} * \left[P_X(ju)\frac{2\pi}{Q} \sum_{k=-\infty}^{\infty} \delta(u - ku_o) \right] \qquad (2.45)$$

$$= \frac{\sin\left(u\frac{Q}{2}\right)}{u\frac{Q}{2}} * \left[\sum_{k=-\infty}^{\infty} P_X(jku_o)\delta(u - ku_o) \right] \qquad (2.46)$$

$$= \sum_{k=-\infty}^{\infty} P_X(jku_o) \left[\frac{\sin\left(u\frac{Q}{2}\right)}{u\frac{Q}{2}} * \delta(u - ku_o) \right] \qquad (2.47)$$

$$P_E(ju) = \sum_{k=-\infty}^{\infty} P_X(jku_o) \frac{\sin\left[(u - ku_o)\frac{Q}{2}\right]}{(u - ku_o)\frac{Q}{2}}. \qquad (2.48)$$

If the quantization theorem is satisfied, i.e. if $P_X(ju) = 0$ for $u > u_o/2$, then there is only one non-zero term ($k = 0$ in (2.48)). The characteristic function of the quantization error is reduced, with $P_X(0) = 1$, to

$$P_E(ju) = \frac{\sin\left(u\frac{Q}{2}\right)}{u\frac{Q}{2}}. \qquad (2.49)$$

Hence, the PDF of the quantization error is

$$p_E(e) = \frac{1}{Q} \mathrm{rect}\left(\frac{e}{Q}\right). \tag{2.50}$$

Sripad and Snyder [Sri77] have modified the sufficient condition of Widrow (band-limited characteristic function of input) for a quantization error of uniform PDF by the weaker condition

$$\boxed{P_X(jku_o) = P_X\left(j\frac{2\pi k}{Q}\right) = 0 \quad \text{for all } k \neq 0.} \tag{2.51}$$

The uniform distribution of the input PDF,

$$p_X(x) = \frac{1}{Q} \mathrm{rect}\left(\frac{x}{Q}\right), \tag{2.52}$$

with characteristic function

$$P_X(ju) = \frac{\sin\left(u\frac{Q}{2}\right)}{u\frac{Q}{2}}, \tag{2.53}$$

does not satisfy Widrow's condition for a band-limited characteristic function, but instead the weaker condition,

$$P_X\left(j\frac{2\pi k}{Q}\right) = \frac{\sin(\pi k)}{\pi k} = 0, \quad \text{for all } k \neq 0, \tag{2.54}$$

is fulfilled. From this follows the uniform PDF (2.49) of the quantization error. The weaker condition from Sripad and Snyder extends the class of input signals for which a uniform PDF of the quantization error can be assumed.

In order to show the deviation from the uniform PDF of the quantization error as a function of the PDF of the input, (2.48) can be written as

$$
\begin{aligned}
P_E(ju) &= P_X(0)\frac{\sin\left[u\frac{Q}{2}\right]}{u\frac{Q}{2}} + \sum_{k=-\infty,k\neq0}^{\infty} P_X\left(j\frac{2\pi k}{Q}\right)\frac{\sin\left[(u-ku_o)\frac{Q}{2}\right]}{(u-ku_o)\frac{Q}{2}} \\
&= \frac{\sin\left[u\frac{Q}{2}\right]}{u\frac{Q}{2}} + \sum_{k=-\infty,k\neq0}^{\infty} P_X\left(j\frac{2\pi k}{Q}\right)\frac{\sin\left[u\frac{Q}{2}\right]}{u\frac{Q}{2}} * \delta(u-ku_0).
\end{aligned} \tag{2.55}
$$

The inverse Fourier transform yields

$$p_E(e) = \frac{1}{Q}\mathrm{rect}\left(\frac{e}{Q}\right)\left[1 + \sum_{k=-\infty,k\neq0}^{\infty} P_X\left(j\frac{2\pi k}{Q}\right)\exp\left(j\frac{2\pi ke}{Q}\right)\right] \tag{2.56}$$

$$= \begin{cases} \frac{1}{Q}\left[1 + \sum_{k\neq0}^{\infty} P_X\left(j\frac{2\pi k}{Q}\right)\exp\left(j\frac{2\pi ke}{Q}\right)\right], & -\frac{Q}{2} \leq e < \frac{Q}{2}, \\ 0, & \text{elsewhere.} \end{cases} \tag{2.57}$$

Equation (2.56) shows the effect of the input PDF on the deviation from a uniform PDF.

Second-order Statistics of Quantization Error

To describe the spectral properties of the error signal, two values E_1 (at time n_1) and E_2 (at time n_2) are considered [Lip92]. The joint PDF is given by

$$p_{E_1 E_2}(e_1, e_2) = \text{rect}\left(\frac{e_1}{Q}, \frac{e_2}{Q}\right)[p_{X_1 X_2}(e_1, e_2) * \delta_{QQ}(e_1, e_2)]. \qquad (2.58)$$

Here $\delta_{QQ}(e_1, e_2) = \delta_Q(e_1) \cdot \delta_Q(e_2)$ and $\text{rect}(e_1/Q, e_2/Q) = \text{rect}(e_1/Q) \cdot \text{rect}(e_2/Q)$. For the Fourier transform of the joint PDF, a similar procedure to that shown by (2.45)–(2.48) leads to

$$P_{E_1 E_2}(ju_1, ju_2) = \sum_{k_1=-\infty}^{\infty} \sum_{k_2=-\infty}^{\infty} P_{X_1 X_2}(jk_1 u_o, jk_2 u_o)$$

$$\cdot \frac{\sin\left[(u_1 - k_1 u_o)\frac{Q}{2}\right]}{(u_1 - k_1 u_o)\frac{Q}{2}} \frac{\sin\left[(u_2 - k_2 u_o)\frac{Q}{2}\right]}{(u_2 - k_2 u_o)\frac{Q}{2}}. \qquad (2.59)$$

If the quantization theorem and/or the Sripad–Snyder condition

$$\boxed{P_{X_1 X_2}(jk_1 u_o, jk_2 u_o) = 0 \quad \text{for all } k_1, k_2 \neq 0} \qquad (2.60)$$

are satisfied then

$$P_{E_1 E_2}(ju_1, ju_2) = \frac{\sin\left[u_1 \frac{Q}{2}\right]}{u_1 \frac{Q}{2}} \frac{\sin\left[u_2 \frac{Q}{2}\right]}{u_2 \frac{Q}{2}}. \qquad (2.61)$$

Thus, for the joint PDF of the quantization error,

$$p_{E_1 E_2}(e_1, e_2) = \frac{1}{Q}\text{rect}\left(\frac{e_1}{Q}\right) \cdot \frac{1}{Q}\text{rect}\left(\frac{e_2}{Q}\right), \quad -\frac{Q}{2} \leq e_1, e_2 < \frac{Q}{2} \qquad (2.62)$$

$$= p_{E_1}(e_1) \cdot p_{E_2}(e_2). \qquad (2.63)$$

Due to the statistical independence of quantization errors (2.63),

$$E\{E_1^m E_2^n\} = E\{E_1^m\} \cdot E\{E_2^n\}. \qquad (2.64)$$

For the moments of the joint PDF,

$$E\{E_1^m E_2^n\} = (-j)^{m+n} \frac{\partial^{m+n}}{\partial u_1^m \partial u_2^n} P_{E_1 E_2}(u_1, u_2)\bigg|_{u_1=0, u_2=0}. \qquad (2.65)$$

From this, it follows for the autocorrelation function with $m = n_2 - n_1$ that

$$r_{EE}(m) = E\{E_1 E_2\} = \begin{cases} E\{E^2\}, & m = 0, \\ E\{E_1 E_2\}, & \text{elsewhere} \end{cases} \qquad (2.66)$$

$$= \begin{cases} \dfrac{Q^2}{12}, & m = 0, \\ 0, & \text{elsewhere} \end{cases} \qquad (2.67)$$

$$= \underbrace{\frac{Q^2}{12}}_{\sigma_E^2} \delta(m). \qquad (2.68)$$

The power density spectrum of the quantization error is then given by

$$S_{EE}(e^{j\Omega}) = \sum_{m=-\infty}^{+\infty} r_{EE}(m)\, e^{-j\Omega m} = \frac{Q^2}{12}, \tag{2.69}$$

which is equal to the variance $\sigma_E^2 = Q^2/12$ of the quantization error (see Fig. 2.15).

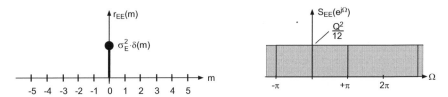

Figure 2.15 Autocorrelation $r_{EE}(m)$ and power density spectrum $S_{EE}(e^{j\Omega})$ of quantization error $e(n)$.

Correlation of Signal and Quantization Error

To describe the correlation of the signal and the quantization error [Sri77], the second moment of the output with (2.26) is derived as follows:

$$E\{Y^2\} = (-j)^2 \frac{d^2 P_Y(ju)}{du^2}\bigg|_{u=0} \tag{2.70}$$

$$= (-j)^2 \sum_{k=-\infty}^{\infty} \left[\ddot{P}_X\left(-\frac{2\pi k}{Q}\right) \frac{\sin(\pi k)}{\pi k} + Q\dot{P}_X\left(-\frac{2\pi k}{Q}\right) \frac{\sin(\pi k) - \pi k \cos(\pi k)}{\pi^2 k^2} \right.$$

$$\left. + \frac{Q^2}{4} P_X\left(-\frac{2\pi k}{Q}\right) \frac{(2 - \pi^2 k^2)\sin(\pi k) - 2\pi k \cos(\pi k)}{\pi^3 k^3} \right] \tag{2.71}$$

$$= E\{X^2\} + \frac{Q}{\pi} \sum_{k=-\infty, k\neq 0}^{\infty} \frac{(-1)^k}{k} \dot{P}_X\left(-\frac{2\pi k}{Q}\right) + E\{E^2\}. \tag{2.72}$$

With the quantization error $e(n) = y(n) - x(n)$,

$$E\{Y^2\} = E\{X^2\} + 2E\{X \cdot E\} + E\{E^2\}, \tag{2.73}$$

where the term $E\{X \cdot E\}$, with (2.72), is written as

$$E\{X \cdot E\} = \frac{Q}{2\pi} \sum_{k=-\infty, k\neq 0}^{\infty} \frac{(-1)^k}{k} \dot{P}_X\left(-\frac{2\pi k}{Q}\right). \tag{2.74}$$

With the assumption of a Gaussian PDF of the input we obtain

$$p_X(x) = \frac{1}{\sqrt{2\pi}\sigma} \exp\left(\frac{-x^2}{2\sigma^2}\right) \tag{2.75}$$

with the characteristic function

$$P_X(ju) = \exp\left(\frac{-u^2\sigma^2}{2}\right). \tag{2.76}$$

Using (2.57), the PDF of the quantization error is then given by

$$p_E(e) = \begin{cases} \dfrac{1}{Q}\left[1 + 2\displaystyle\sum_{k=1}^{\infty}\cos\left(\dfrac{2\pi ke}{Q}\right)\exp\left(-\dfrac{2\pi^2 k^2\sigma^2}{Q^2}\right)\right], & -\dfrac{Q}{2} \le e < \dfrac{Q}{2}, \\ 0, & \text{elsewhere.} \end{cases} \tag{2.77}$$

Figure 2.16a shows the PDF (2.77) of the quantization error for different variances of the input.

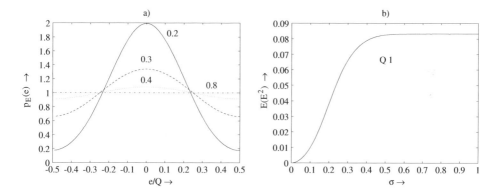

Figure 2.16 (a) PDF of quantization error for different standard deviations of a Gaussian PDF input. (b) Variance of quantization error for different standard deviations of a Gaussian PDF input.

For the mean value and the variance of a quantization error, it follows using (2.77) that $E\{E\} = 0$ and

$$E\{E^2\} = \int_{-\infty}^{\infty} e^2 p_E(e)\,de = \frac{Q^2}{12}\left[1 + \frac{12}{\pi^2}\sum_{k=1}^{\infty}\frac{(-1)^k}{k^2}\exp\left(-\frac{2\pi^2 k^2\sigma^2}{Q^2}\right)\right]. \tag{2.78}$$

Figure 2.16b shows the variance of the quantization error (2.78) for different variances of the input.

For a Gaussian PDF input as given by (2.75) and (2.76), the correlation (see (2.74)) between input and quantization error is expressed as

$$E\{X \cdot E\} = 2\sigma^2 \sum_{k=1}^{\infty}(-1)^k \exp\left(-\frac{2\pi^2 k^2\sigma^2}{Q^2}\right). \tag{2.79}$$

The correlation is negligible for large values of σ/Q.

2.2   Dither

2.2.1   Basics

The *requantization* (renewed quantization of already quantized signals) to limited word-lengths occurs repeatedly during storage, format conversion and signal processing algorithms. Here, small signal levels lead to error signals which depend on the input. Owing to quantization, nonlinear distortion occurs for low-level signals. The conditions for the classical quantization model are not satisfied anymore. To reduce these effects for signals of small amplitude, a linearization of the nonlinear characteristic curve of the quantizer is performed. This is done by adding a random sequence $d(n)$ to the quantized signal $x(n)$ (see Fig. 2.17) before the actual quantization process. The specification of the word-length is shown in Fig. 2.18. This random signal is called *dither*. The statistical independence of the error signal from the input is not achieved, but the conditional moments of the error signal can be affected [Lip92, Ger89, Wan92, Wan00].

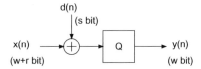

Figure 2.17 Addition of a random sequence before a quantizer.

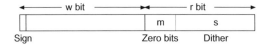

Figure 2.18 Specification of the word-length.

The sequence $d(n)$, with amplitude range $(-Q/2 \leq d(n) \leq Q/2)$, is generated with the help of a random number generator and is added to the input. For a dither value with $Q = 2^{-(w-1)}$:

$$d_k = k2^{-r}Q, \quad -2^{s-1} \leq k \leq 2^{s-1} - 1. \tag{2.80}$$

The index k of the random number d_k characterizes the value from the set of $N = 2^s$ possible numbers with the probability

$$P(d_k) = \begin{cases} 2^{-s}, & -2^{s-1} \leq k \leq 2^{s-1} - 1, \\ 0, & \text{elsewhere.} \end{cases} \tag{2.81}$$

With the mean value $\overline{d} = \sum_k d_k P(d_k)$, the variance $\sigma_D^2 = \sum_k [d_k - \overline{d}]^2 P(d_k)$ and the quadratic mean $\overline{d^2} = \sum_k d_k^2 P(d_k)$, we can rewrite the variance as $\sigma_d^2 = \overline{d^2} - \overline{d}^2$.

For a static input amplitude V and the dither value d_k the rounding operation [Lip86] is expressed as

$$g(V + d_k) = Q \left\lfloor \frac{V + d_k}{Q} + 0.5 \right\rfloor. \tag{2.82}$$

For the mean of the output $\overline{g}(V)$ as a function of the input V, we can write

$$\overline{g}(V) = \sum_k g(V + d_k) P(d_k). \tag{2.83}$$

The quadratic mean of the output $\overline{g^2}(V)$ for input V is given by

$$\overline{g^2}(V) = \sum_k g^2(V + d_k) P(d_k). \tag{2.84}$$

For the variance $d_R^2(V)$ for input V,

$$d_R^2(V) = \sum_k \{g(V + d_k) - \overline{g}(V)\}^2 P(d_k) = \overline{g^2}(V) - \{\overline{g}(V)\}^2. \tag{2.85}$$

The above equations have the input V as a parameter. Figures 2.19 and 2.20 illustrate the mean output $\overline{g}(V)$ and the standard deviation $d_R(V)$ within a quantization step size, given by (2.83), (2.84) and (2.85). The examples of rounding and truncation demonstrate the linearization of the characteristic curve of the quantizer. The coarse step size is replaced by a finer one. The quadratic deviation from the mean output $d_R^2(V)$ is termed *noise modulation*. For a uniform PDF dither, this noise modulation depends on the amplitude (see Figs. 2.19 and 2.20). It is maximum in the middle of the quantization step size and approaches zero toward the end. The linearization and the suppression of the noise modulation can be achieved by a triangular PDF dither with bipolar characteristic [Van89] and rounding operation (see Fig. 2.20). Triangular PDF dither is obtained by adding two statistically independent dither signals with uniform PDF (convolution of PDFs). A dither signal with higher-order PDF is not necessary for audio signals [Lip92, Wan00].

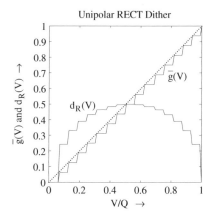
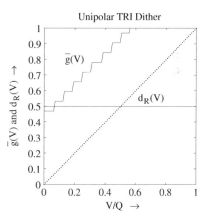

Figure 2.19 Truncation – linearizing and suppression of noise modulation ($s = 4$, $m = 0$).

The total noise power for this quantization technique consists of the dither power and the power of the quantization error [Lip86]. The following noise powers are obtained by integration with respect to V.

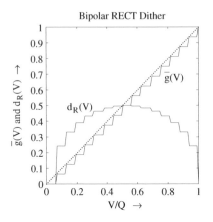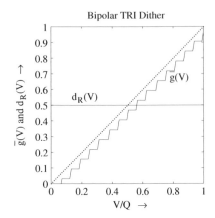

Figure 2.20 Rounding – linearizing and suppression of noise modulation ($s = 4$, $m = 1$).

1. Mean dither power d^2:

$$d^2 = \frac{1}{Q} \int_0^Q d_R^2(V)\, dV \tag{2.86}$$

$$= \frac{1}{Q} \int_0^Q \sum_k \{g(V + d_k) - \overline{g}(V)\}^2 P(d_k)\, dV. \tag{2.87}$$

(This is equal to the deviation from mean output in accordance with (2.83).)

2. Mean of total noise power d_{tot}^2:

$$d_{\text{tot}}^2 = \frac{1}{Q} \int_0^Q \sum_k \{g(V + d_k) - V\}^2 P(d_k)\, dV. \tag{2.88}$$

(This is equal to the deviation from an ideal straight line.)

In order to derive a relationship between d_{tot}^2 and d^2, the quantization error given by

$$Q(V + d_k) = g(V + d_k) - (V + d_k) \tag{2.89}$$

is used to rewrite (2.88) as

$$d_{\text{tot}}^2 = \sum_k P(d_k) \frac{1}{Q} \int_0^Q (Q^2(V + d_k) + 2d_k Q(V + d_k) + d_k^2)\, dV \tag{2.90}$$

$$= \sum_k P(d_k) \frac{1}{Q} \int_0^Q Q^2(V + d_k)\, dV + 2 \sum_k d_k P(d_k) \frac{1}{Q} \int_0^Q Q(V + d_k)\, dV$$

$$+ \sum_k d_k^2 P(d_k) \frac{1}{Q} \int_0^Q dV. \tag{2.91}$$

The integrals in (2.91) are independent of d_k. Moreover, $\sum_k P(d_k) = 1$. With the mean value of the quantization error

$$\bar{e} = \frac{1}{Q} \int_0^Q Q(V)\, dV \qquad (2.92)$$

and the quadratic mean error

$$\overline{e^2} = \frac{1}{Q} \int_0^Q Q^2(V)\, dV, \qquad (2.93)$$

it is possible to rewrite (2.91) as

$$d_{\text{tot}}^2 = \overline{e^2} + 2\overline{d}\bar{e} + \overline{d^2}. \qquad (2.94)$$

With $\sigma_E^2 = \overline{e^2} - \bar{e}^2$ and $\sigma_D^2 = \overline{d^2} - \bar{d}^2$, (2.94) can be written as

$$\boxed{d_{\text{tot}}^2 = \sigma_E^2 + (\bar{d} + \bar{e})^2 + \sigma_D^2.} \qquad (2.95)$$

Equations (2.94) and (2.95) describe the total noise power as a function of the quantization ($\bar{e}$, $\overline{e^2}$, σ_E^2) and the dither addition ($\bar{d}$, $\overline{d^2}$, σ_D^2). It can be seen that for zero-mean quantization, the middle term in (2.95) results in $\bar{d} + \bar{e} = 0$. The acoustically perceptible part of the total error power is represented by σ_e^2 and σ_d^2.

2.2.2 Implementation

The random sequence $d(n)$ is generated with the help of a random number generator with uniform PDF. For generating a triangular PDF random sequence, two independent uniform PDF random sequences $d_1(n)$ and $d_2(n)$ can be added. In order to generate a triangular high-pass dither, the dither value $d_1(n)$ is added to $-d_1(n-1)$. Thus, only one random number generator is required. In conclusion, the following dither sequences can be implemented:

$$d_{\text{RECT}}(n) = d_1(n), \qquad (2.96)$$
$$d_{\text{TRI}}(n) = d_1(n) + d_2(n), \qquad (2.97)$$
$$d_{\text{HP}}(n) = d_1(n) - d_1(n-1). \qquad (2.98)$$

The power density spectra of triangular PDF dither and triangular PDF HP dither are shown in Fig. 2.21. Figure 2.22 shows histograms of a uniform PDF dither and a triangular PDF high-pass dither together with their respective power density spectra. The amplitude range of a uniform PDF dither lies between $\pm Q/2$, whereas it lies between $\pm Q$ for triangular PDF dither. The total noise power for triangular PDF dither is doubled.

2.2.3 Examples

The effect of the input amplitude of the quantizer is shown in Fig. 2.23 for a 16-bit quantizer ($Q = 2^{-15}$). A quantized sinusoidal signal with amplitude 2^{-15} (1-bit amplitude)

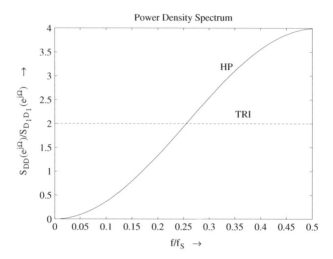

Figure 2.21 Normalized power density spectrum for triangular PDF dither (TRI) with $d_1(n) + d_2(n)$ and triangular PDF high-pass dither (HP) with $d_1(n) - d_1(n-1)$.

and frequency $f/f_S = 64/1024$ is shown in Fig. 2.23a,b for rounding and truncation. Figure 2.23c,d shows their corresponding spectra. For truncation, Fig. 2.23c shows the spectral line of the signal and the spectral distribution of the quantization error with the harmonics of the input signal. For rounding (Fig. 2.23d with special signal frequency $f/f_S = 64/1024$), the quantization error is concentrated in uneven harmonics.

In the following, only the rounding operation is used. By adding a uniform PDF random signal to the actual signal before quantization, the quantized signal shown in Fig. 2.24a results. The corresponding power density spectrum is illustrated in Fig. 2.24c. In the time domain, it is observed that the 1-bit amplitudes approach zero so that the regular pattern of the quantized signal is affected. The resulting power density spectrum in Fig. 2.24c shows that the harmonics do not occur anymore and the noise power is uniformly distributed over the frequencies. For triangular PDF dither, the quantized signal is shown in Fig. 2.24b. Owing to triangular PDF, amplitudes $\pm 2Q$ occur besides the signal values $\pm Q$ and zero. Figure 2.24d shows the increase of the total noise power.

In order to illustrate the noise modulation for uniform PDF dither, the amplitude of the input is reduced to $A = 2^{-18}$ and the frequency is chosen as $f/f_S = 14/1024$. This means that input amplitude to the quantizer is 0.25 bit. For a quantizer without additive dither, the quantized output signal is zero. For RECT dither, the quantized signal is shown in Fig. 2.25a. The input signal with amplitude $0.25Q$ is also shown. The power density spectrum of the quantized signal is shown in Fig. 2.25c. The spectral line of the signal and the uniform distribution of the quantization error can be seen. But in the time domain, a correlation between positive and negative amplitudes of the input and the quantized positive and negative values of the output can be observed. In hearing tests this noise

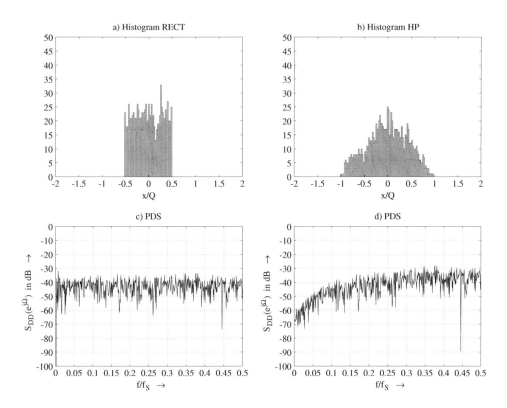

Figure 2.22 (a,b) Histogram and (c,d) power density spectrum of uniform PDF dither (RECT) with $d_1(n)$ and triangular PDF high-pass dither (HP) with $d_1(n) - d_1(n-1)$.

modulation occurs if the amplitude of the input is decreased continuously and falls below the amplitude of the quantization step. This process occurs for all fade-out processes that occur in speech and music signals. For positive low-amplitude signals, two output states, zero and Q, occur, and for negative low-amplitude signals, the output states zero and −Q, occur. This is observed as a disturbing rattle which is overlapped to the actual signal. If the input level is further reduced the quantized output approaches zero.

In order to reduce this noise modulation at low levels, a triangular PDF dither is used. Figure 2.25b shows the quantized signal and Fig. 2.25d shows the power density spectrum. It can be observed that the quantized signal has an irregular pattern. Hence a direct association of positive half-waves with the positive output values as well as vice versa is not possible. The power density spectrum shows that spectral line of the signal along with an increase in noise power owing to triangular PDF dither. In acoustic hearing tests, the use of triangular PDF dither results in a constant noise floor even if the input level is reduced to zero.

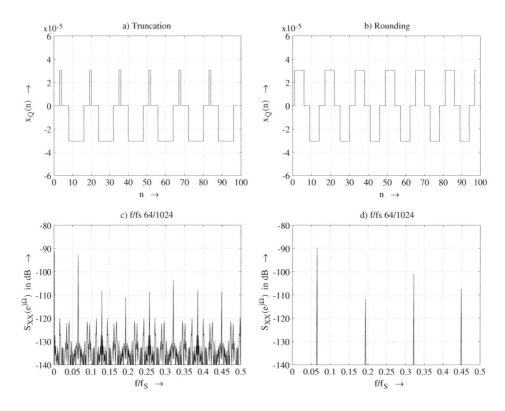

Figure 2.23 One-bit amplitude – quantizer with truncation (a,c) and rounding (b,d).

2.3   Spectrum Shaping of Quantization – Noise Shaping

Using the linear model of a quantizer in Fig. 2.26 and the relations

$$e(n) = y(n) - x(n), \tag{2.99}$$

$$y(n) = [x(n)]_Q \tag{2.100}$$

$$= x(n) + e(n), \tag{2.101}$$

the quantization error $e(n)$ may be isolated and fed back through a transfer function $H(z)$ as shown in Fig. 2.27. This leads to the spectral shaping of the quantization error as given by

$$y(n) = [x(n) - e(n) * h(n)]_Q \tag{2.102}$$

$$= x(n) + e(n) - e(n) * h(n), \tag{2.103}$$

$$e_1(n) = y(n) - x(n) \tag{2.104}$$

$$= e(n) * (\delta(n) - h(n)), \tag{2.105}$$

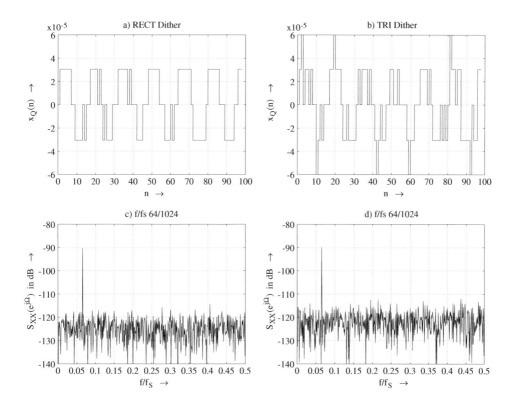

Figure 2.24 One-bit amplitude – rounding with RECT dither (a,c) and TRI dither (b,d).

and the corresponding Z-transforms

$$Y(z) = X(z) + E(z)(1 - H(z)) \tag{2.106}$$

$$E_1(z) = E(z)(1 - H(z)). \tag{2.107}$$

A simple spectrum shaping of the quantization error $e(n)$ is achieved by feeding back with $H(z) = z^{-1}$ as shown in Fig. 2.28, and leads to

$$y(n) = [x(n) - e(n-1)]_Q \tag{2.108}$$

$$= x(n) - e(n-1) + e(n), \tag{2.109}$$

$$e_1(n) = y(n) - x(n) \tag{2.110}$$

$$= e(n) - e(n-1), \tag{2.111}$$

and the Z-transforms

$$Y(z) = X(z) + E(z)(1 - z^{-1}), \tag{2.112}$$

$$E_1(z) = E(z)(1 - z^{-1}). \tag{2.113}$$

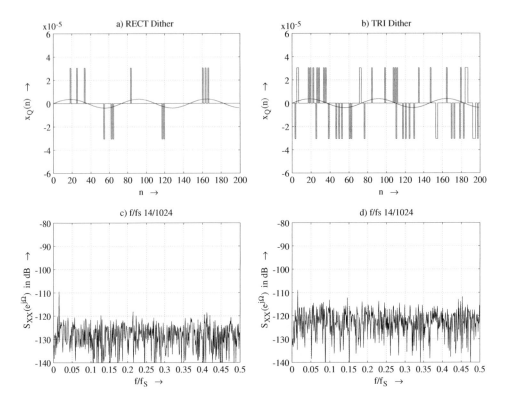

Figure 2.25 0.25-bit amplitude – rounding with RECT dither (a,c) and TRI dither (b,d).

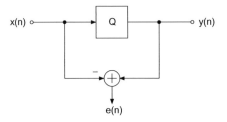

Figure 2.26 Linear model of quantizer.

Equation (2.113) shows a high-pass weighting of the original error signal $e(n)$. By choosing $H(z) = z^{-1}(-2 + z^{-1})$, second-order high-pass weighting given by

$$E_2(z) = E(z)(1 - 2z^{-1} + z^{-2}) \tag{2.114}$$

can be achieved. The power density spectrum of the error signal for the two cases is given by

$$S_{E_1 E_1}(e^{j\Omega}) = |1 - e^{-j\Omega}|^2 S_{EE}(e^{j\Omega}), \tag{2.115}$$

$$S_{E_2 E_2}(e^{j\Omega}) = |1 - 2e^{-j\Omega} + e^{-j2\Omega}|^2 S_{EE}(e^{j\Omega}). \tag{2.116}$$

Figure 2.29 shows the weighting of power density spectrum by this noise shaping technique.

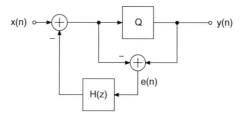

Figure 2.27 Spectrum shaping of quantization error.

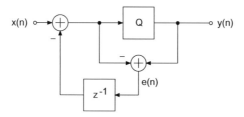

Figure 2.28 High-pass spectrum shaping of quantization error.

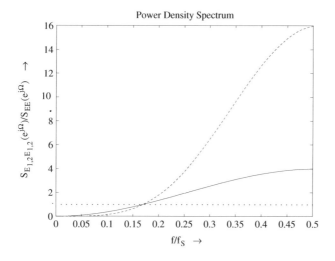

Figure 2.29 Spectrum shaping $(S_{EE}(e^{j\Omega}) \cdots, S_{E_1 E_1}(e^{j\Omega}) —, S_{E_2 E_2}(e^{j\Omega}) - - -)$.

By adding a dither signal $d(n)$ (see Fig. 2.30), the output and the error are given by

$$y(n) = [x(n) + d(n) - e(n-1)]_Q \qquad (2.117)$$
$$= x(n) + d(n) - e(n-1) + e(n) \qquad (2.118)$$

and

$$e_1(n) = y(n) - x(n) \tag{2.119}$$
$$= d(n) + e(n) - e(n-1). \tag{2.120}$$

For the Z-transforms we write

$$Y(z) = X(z) + E(z)(1 - z^{-1}) + D(z), \tag{2.121}$$
$$E_1(z) = E(z)(1 - z^{-1}) + D(z). \tag{2.122}$$

The modified error signal $e_1(n)$ consists of the dither and the high-pass shaped quantization error.

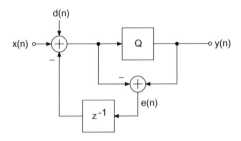

Figure 2.30 Dither and spectrum shaping.

By moving the addition (Fig. 2.31) of the dither directly before the quantizer, a high-pass spectrum shaping is obtained for both the error signal and the dither. Here the following relationships hold:

$$y(n) = [x(n) + d(n) - e_0(n-1)]_Q \tag{2.123}$$
$$= x(n) + d(n) - e_0(n-1) + e(n), \tag{2.124}$$
$$e_0(n) = y(n) - (x(n) - e_0(n-1)) \tag{2.125}$$
$$= d(n) + e(n), \tag{2.126}$$
$$y(n) = x(n) + d(n) - d(n-1) + e(n) - e(n-1), \tag{2.127}$$
$$e_1(n) = d(n) - d(n-1) + e(n) - e(n-1), \tag{2.128}$$

with the Z-transforms given by

$$Y(z) = X(z) + E(z)(1 - z^{-1}) + D(z)(1 - z^{-1}), \tag{2.129}$$
$$E_1(z) = E(z)(1 - z^{-1}) + D(z)(1 - z^{-1}). \tag{2.130}$$

Apart from the discussed feedback structures which are easy to implement on a digital signal processor and which lead to high-pass noise shaping, psychoacoustic-based noise shaping methods have been proposed in the literature [Ger89, Wan92, Hel07]. These methods use special approximations of the hearing threshold (threshold in quiet, absolute threshold) for the feedback structure $1 - H(z)$. Figure 2.32a shows several hearing threshold models as a function of frequency [ISO389, Ter79, Wan92]. It can be seen that the sensitivity of human hearing is high for frequencies between 2 and 6 kHz and sharply decreases for high and low frequencies. Figure 2.32b also shows the inverse ISO 389-7 threshold curve which represents an approximation of the filtering operation in our perception. The feedback filter of the noise shaper should affect the quantization error with the inverse ISO 389 weighting curve. Hence, the noise power in the frequency range with high sensitivity should be reduced and shifted toward lower and higher frequencies. Figure 2.33a shows the unweighted power density spectra of the quantization error for three special filters $H(z)$ [Wan92, Hel07]. Figure 2.33b depicts the same three power density spectra, weighted by the inverse ISO 389 threshold of Fig. 2.32b. These weighted power density spectra show that the perceived noise power is reduced by all three noise shapers versus the frequency axis. Figure 2.34 shows a sinusoid with amplitude $Q = 2^{-15}$, which is quantized to $w = 16$ bits with psychoacoustic noise shaping. The quantized signal $x_Q(n)$ consists of different amplitudes reflecting the low-level signal. The power density spectrum of the quantized signal reflects the psychoacoustic weighting of the noise shaper with a fixed filter. A time-variant psychoacoustic noise shaping is described in [DeK03, Hel07], where the instantaneous masking threshold is used for adaptation of a time-variant filter.

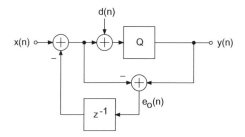

Figure 2.31 Modified dither and spectrum shaping.

2.4 Number Representation

The different applications in digital signal processing and transmission of audio signals leads to the question of the type of number representation for digital audio signals. In this section, basic properties of fixed-point and floating-point number representation in the context of digital audio signal processing are presented.

2.4.1 Fixed-point Number Representation

In general, an arbitrary real number x can be approximated by a finite summation

$$x_Q = \sum_{i=0}^{w-1} b_i 2^i, \tag{2.131}$$

where the possible values for b_i are 0 and 1.

The fixed-point number representation with a finite number w of binary places leads to four different interpretations of the number range (see Table 2.1 and Fig. 2.35).

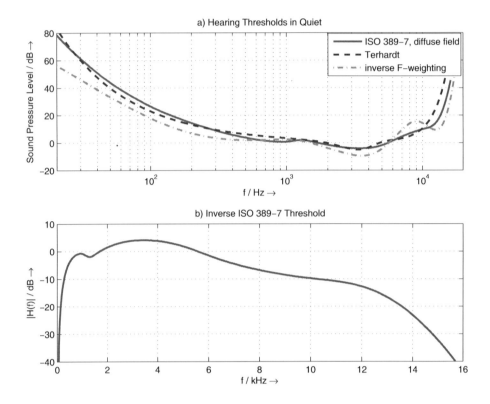

Figure 2.32 (a) Hearing thresholds in quiet. (b) Inverse ISO 389-7 threshold curve.

Table 2.1 Bit location and range of values.

Type	Bit location	Range of values
Signed 2's c.	$x_Q = -b_0 + \sum_{i=1}^{w-1} b_{-i} 2^{-i}$	$-1 \le x_Q \le 1 - 2^{-(w-1)}$
Unsigned 2's c.	$x_Q = \sum_{i=1}^{w} b_{-i} 2^{-i}$	$0 \le x_Q \le 1 - 2^{-w}$
Signed int.	$x_Q = -b_{w-1} 2^{w-1} + \sum_{i=0}^{w-2} b_i 2^i$	$-2^{w-1} \le x_Q \le 2^{w-1} - 1$
Unsigned int.	$x_Q = \sum_{i=0}^{w-1} b_i 2^i$	$0 \le x_Q \le 2^w - 1$

The signed fractional representation (2's complement) is the usual format for digital audio signals and for algorithms in fixed-point arithmetic. For address and modulo operation, the unsigned integer is used. Owing to finite word-length w, overflow occurs as shown in Fig. 2.36. These curves have to be taken into consideration while carrying out operations, especially additions in 2's complement arithmetic.

Quantization is carried out with techniques as shown in Table 2.2 for rounding and truncation. The quantization step size is characterized by $Q = 2^{-(w-1)}$ and the symbol $\lfloor x \rfloor$ denotes the biggest integer smaller than or equal to x. Figure 2.37 shows the rounding and

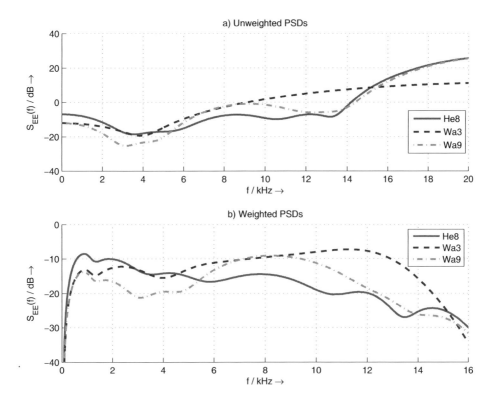

Figure 2.33 Power density spectrum of three filter approximations (Wa3 third-order filter, Wa9 ninth-order filter, He8 eighth-order filter [Wan92, Hel07]): (a) unweighted PSDs, (b) inverse ISO 389-7 weighted PSDs.

truncation curves for 2's complement number representation. The absolute error shown in Fig. 2.37 is given by $e = x_Q - x$.

Table 2.2 Rounding and truncation of 2s complement numbers.

Type	Quantization	Error limits
2's c. (r)	$x_Q = Q\lfloor Q^{-1}x + 0.5 \rfloor$	$-Q/2 \leq x_Q - x \leq Q/2$
2's c. (t)	$x_Q = Q\lfloor Q^{-1}x \rfloor$	$-Q \leq x_Q - x \leq 0$

Digital audio signals are coded in the 2's complement number representation. For 2's complement representation, the range of values from $-X_{\max}$ to $+X_{\max}$ is normalized to the range -1 to $+1$ and is represented by the weighted finite sum $x_Q = -b_0 + b_1 \cdot 0.5 + b_2 \cdot 0.25 + b_3 \cdot 0.125 + \cdots + b_{w-1} \cdot 2^{-(w-1)}$. The variables b_0 to b_{w-1} are called bits and can take the values 1 or 0. The bit b_0 is called the MSB (most significant bit) and b_{w-1} is called the LSB (least significant bit). For positive numbers, b_0 is equal to 0 and for

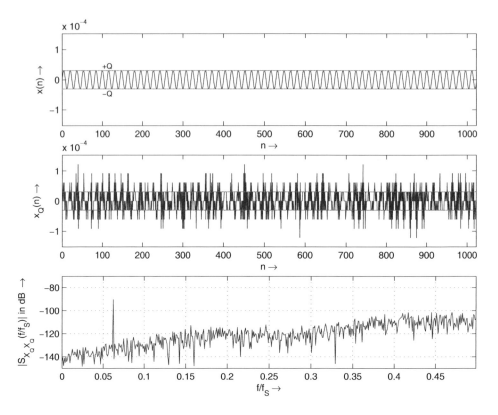

Figure 2.34 Psychoacoustic noise shaping: signal $x(n)$, quantized signal $x_Q(n)$ and power density spectrum of quantized signal.

Bit	31	30	29				2	1	0	
	-2^0	2^{-1}	2^{-2}	.	.	.	2^{-29}	2^{-30}	2^{-31}	Signed 2's Complement
	2^{-1}	2^{-2}	2^{-3}	.	.	.	2^{-30}	2^{-31}	2^{-32}	Unsigned 2's Complement
	-2^{31}	2^{30}	2^{29}	.	.	.	2^2	2^1	2^0	Signed Integer
	2^{31}	2^{30}	2^{29}	.	.	.	2^2	2^1	2^0	Unsigned Integer

Figure 2.35 Fixed-point formats.

negative numbers b_0 equals 1. For a 3-bit quantization (see Fig. 2.38), a quantized value can be represented by $x_Q = -b_0 + b_1 \cdot 0.5 + b_2 \cdot 0.25$. The smallest quantization step size is 0.25. For a positive number 0.75 it follows that $0.75 = -0 + 1 \cdot 0.5 + 1 \cdot 0.25$. The binary coding for 0.75 is 011.

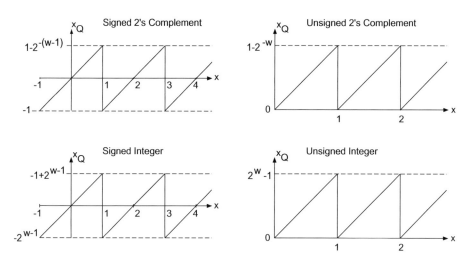

Figure 2.36 Number range.

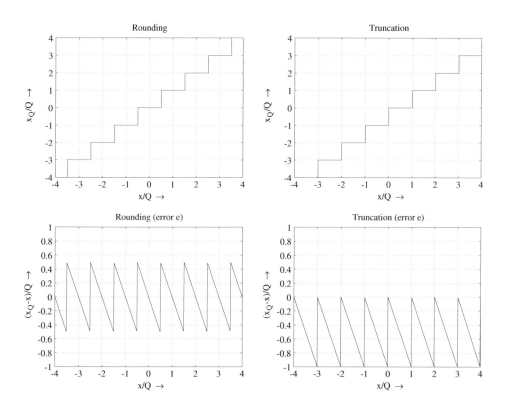

Figure 2.37 Rounding and truncation curves.

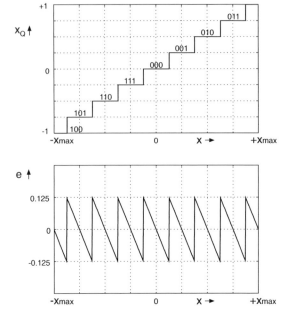

Figure 2.38 Rounding curve and error signal for $w = 3$ bits.

Dynamic Range. The dynamic range of a number representation is defined as the ratio of maximum to minimum number. For fixed-point representation with

$$x_{Q\,\text{max}} = (1 - 2^{-(w-1)}), \tag{2.132}$$

$$x_{Q\,\text{min}} = 2^{-(w-1)}, \tag{2.133}$$

the dynamic range is given by

$$\text{DR}_F = 20 \log_{10} \left(\frac{x_{Q\,\text{max}}}{x_{Q\,\text{min}}} \right) = 20 \log_{10} \left(\frac{1 - Q}{Q} \right)$$

$$= 20 \log_{10} (2^{w-1} - 1) \text{ dB}. \tag{2.134}$$

Multiplication and Addition of Fixed-point Numbers. For the multiplication of two fixed-point numbers in the range from -1 to $+1$, the result is always less than 1. For the addition of two fixed-point numbers, care must be taken for the result to remain in the range from -1 to $+1$. An addition of $0.6 + 0.7 = 1.3$ must be carried out in the form $0.5(0.6 + 0.7) = 0.65$. This multiplication with the factor 0.5 or generally 2^{-s} is called scaling. An integer in the range from 1 to, for instance, 8 is chosen for the scaling coefficient s.

Error Model. The quantization process for fixed-point numbers can be approximated as an the addition of an error signal $e(n)$ to the signal $x(n)$ (see Fig. 2.39). The error signal is a random signal with white power density spectrum.

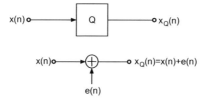

Figure 2.39 Model of a fixed-point quantizer.

Signal-to-noise Ratio. The signal-to-noise ratio for a fixed-point quantizer is defined by

$$\text{SNR} = 10 \log_{10} \left(\frac{\sigma_X^2}{\sigma_E^2} \right), \tag{2.135}$$

where σ_X^2 is the signal power and σ_E^2 is the noise power.

2.4.2 Floating-point Number Representation

The representation of a floating-point number is given by

$$x_Q = M_G \, 2^{E_G} \tag{2.136}$$

with

$$0.5 \leq M_G < 1, \tag{2.137}$$

where M_G denotes the normalized mantissa and E_G the exponent. The normalized standard format (IEEE) is shown in Fig. 2.40 and special cases are given in Table 2.3. The mantissa M is implemented with a word-length of w_M bits and is in fixed-point number representation. The exponent E is implemented with a word-length of w_E bits and is an integer in the range from $-2^{w_E-1} + 2$ to $2^{w_E-1} - 1$. For an exponent word-length of $w_E = 8$ bits, its range of values lies between -126 and $+127$. The range of values of the mantissa lies between 0.5 and 1. This is referred to as the normalized mantissa and is responsible for the unique representation of a number. For a fixed-point number in the range between 0.5 and 1, it follows that the exponent of the floating-point number representation is $E = 0$. To represent a fixed-point number in the range between 0.25 and 0.5 in floating-point form, the range of values of the normalized mantissa M lies between 0.5 and 1, and for the exponent it follows that $E = -1$. As an example, for a fixed-point number 0.75 the floating-point number $0.75 \cdot 2^0$ results. The fixed-point number 0.375 is not represented as the floating-point number $0.375 \cdot 2^0$. With the normalized mantissa, the floating-point number is expressed as $0.75 \cdot 2^{-1}$. Owing to normalization, the ambiguity of floating-point number representation is avoided. Numbers greater than 1 can be represented. For example, 1.5 becomes $0.75 \cdot 2^1$ in floating-point number representation.

Figure 2.41 shows the rounding and truncations curves for floating-point representation and the absolute error $e = x_Q - x$. The curves for floating-point quantization show that for small amplitudes small quantization step sizes occur. In contrast to fixed-point representation, the absolute error is dependent on the input signal.

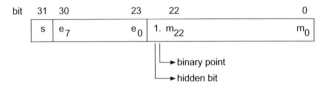

Figure 2.40 Floating-point number representation.

Table 2.3 Special cases of floating-point number representation.

Type	Exponent	Mantissa	Value
NAN	255	$\neq 0$	Undefined
Infinity	255	0	$(-1)^s$ infinity
Normal	$1 \leq e \leq 254$	Any	$(-1)^s (0.m) 2^{e-127}$
Zero	0	0	$(-1)^s \cdot 0$

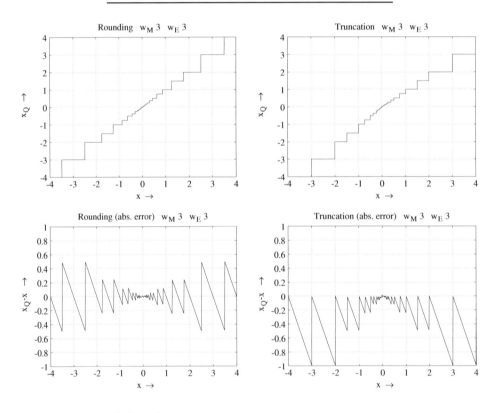

Figure 2.41 Rounding and truncation curves for floating-point representation.

In the interval

$$2^{E_G} \le x < 2^{E_G+1},$$ (2.138)

the quantization step is given by

$$Q_G = 2^{-(w_M-1)} 2^{E_G}.$$ (2.139)

For the relative error

$$e_r = \frac{x_Q - x}{x}$$ (2.140)

of the floating-point representation, a constant upper limit can be stated as

$$|e_r| \le 2^{-(w_M-1)}.$$ (2.141)

Dynamic Range. With the maximum and minimum numbers given by

$$x_{Q\,\text{max}} = (1 - 2^{-(w_M-1)}) 2^{E_G\,\text{max}},$$ (2.142)

$$x_{Q\,\text{min}} = 0.5\, 2^{E_G\,\text{min}},$$ (2.143)

and

$$E_{G\,\text{max}} = 2^{w_E-1} - 1,$$ (2.144)

$$E_{G\,\text{min}} = -2^{w_E-1} + 2,$$ (2.145)

the dynamic range for floating-point representation is given by

$$DR_G = 20 \log_{10}\left(\frac{(1 - 2^{-(w_M-1)}) 2^{E_G\,\text{max}}}{0.5\, 2^{E_G\,\text{min}}} \right)$$

$$= 20 \log_{10}(1 - 2^{-(w_M-1)}) 2^{E_G\,\text{max} - E_G\,\text{min} + 1}$$

$$= 20 \log_{10}(1 - 2^{-(w_M-1)}) 2^{2^{w_E} - 2} \text{ dB.}$$ (2.146)

Multiplication and Addition of Floating-point Numbers. For multiplications with floating-point numbers, the exponents of both numbers $x_{Q1} = M_1 2^{E_1}$ and $x_{Q2} = M_2 2^{E_2}$ are added and the mantissas are multiplied. The resulting exponent $E_G = E_1 + E_2$ is adjusted so that $M_G = M_1 M_2$ lies in the interval $0.5 \le M_G < 1$. For additions the smaller number is denormalized to get the same exponent. Then both mantissa are added and the result is normalized.

Error Model. With the definition of the relative error $e_r(n) = [x_Q(n) - x(n)]/x(n)$ the quantized signal can be written as

$$x_Q(n) = x(n) \cdot (1 + e_r(n)) = x(n) + x(n) \cdot e_r(n).$$ (2.147)

Floating-point quantization can be modeled as an additive error signal $e(n) = x(n) \cdot e_r(n)$ to the signal $x(n)$ (see Fig. 2.42).

Signal-to-noise Ratio. Under the assumption that the relative error is independent of the input x, the noise power of the floating-point quantizer can be written as

$$\sigma_E^2 = \sigma_X^2 \cdot \sigma_{E_r}^2.$$ (2.148)

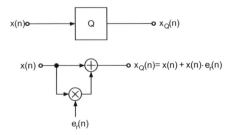

Figure 2.42 Model of a floating-point quantizer.

For the signal-to-noise-ratio, we can derive

$$\text{SNR} = 10 \log_{10}\left(\frac{\sigma_X^2}{\sigma_E^2}\right) = 10 \log_{10}\left(\frac{\sigma_X^2}{\sigma_X^2 \cdot \sigma_{E_r}^2}\right) = 10 \log_{10}\left(\frac{1}{\sigma_{E_r}^2}\right). \tag{2.149}$$

Equation (2.149) shows that the signal-to-noise ratio is independent of the level of the input. It is only dependent on the noise power $\sigma_{E_r}^2$ which, in turn, is only dependent on the word-length w_M of the mantissa of the floating-point representation.

2.4.3   Effects on Format Conversion and Algorithms

First, a comparison of signal-to-noise ratios is made for the fixed-point and floating-point number representation. Figure 2.43 shows the signal-to-noise ratio as a function of input level for both number representations. The fixed-point word-length is $w = 16$ bits. The word-length of the mantissa in floating-point representation is also $w_M = 16$ bits, whereas that of the exponent is $w_E = 4$ bits. The signal-to-noise ratio for floating-point representation shows that it is independent of input level and varies as a sawtooth curve in a 6 dB grid. If the input level is so low that a normalization of the mantissa due to finite number representation is not possible, then the signal-to-noise ratio is comparable to fixed-point representation. While using the full range, both fixed-point and floating-point result in the same signal-to-noise ratio. It can be observed that the signal-to-noise ratio for fixed-point representation depends on the input level. This signal-to-noise ratio in the digital domain is an exact image of the level-dependent signal-to-noise ratio of an analog signal in the analog domain. A floating-point representation cannot improve this signal-to-noise ratio. Rather, the floating-point curve is vertically shifted downwards to the value of signal-to-noise ratio of an analog signal.

AD/DA Conversion. Before processing, storing and transmission of audio signals, the analog audio signal is converted into a digital signal. The precision of this conversion depends on the word-length w of the AD converter. The resulting signal-to-noise ratio is $6w$ dB for uniform PDF inputs. The signal-to-noise ratio in the analog domain depends on the level. This linear dependence of signal-to-noise ratio on level is preserved after AD conversion with subsequent fixed-point representation.

Digital Audio Formats. The basis for established digital audio transmission formats is provided in the previous section on AD/DA conversion. The digital two-channel AES/EBU

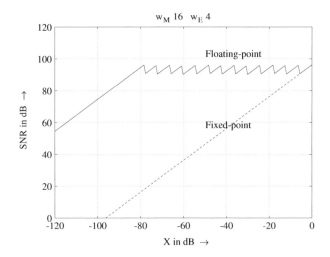

Figure 2.43 Signal-to-noise ratio for an input level.

interface [AES92] and 56-channel MADI interface [AES91] both operate with fixed-point representation with a word-length of at most 24 bits per channel.

Storage and Transmission. Besides the established storage media like compact disc and DAT which were exclusively developed for audio application, there are storage systems like hard discs in computers. These are based on magnetic or magneto-optic principles. The systems operate with fixed-point number representation. With regard to the transmission of digital audio signals for band-limited transmission channels like satellite broadcasting (Digital Satellite Radio, DSR) or terrestrial broadcasting, it is necessary to reduce bit rates. For this, a conversion of a block of linearly coded samples is carried out in a so-called block floating-point representation in DSR. In the context of DAB, a data reduction of linear coded samples is carried out based on psychoacoustic criteria.

Equalizers. While implementing equalizers with recursive digital filters, the signal-to-noise ratio depends on the choice of the recursive filter structure. By a suitable choice of a filter structure and methods to spectrally shape the quantization errors, optimal signal-to-noise ratios are obtained for a given word-length. The signal-to-noise ratio for fixed-point representation depends on the word-length and for floating-point representation on the word-length of the mantissa. For filter implementations with fixed-point arithmetic, boost filters have to be implemented with a scaling within the filter algorithm. The properties of floating-point representation take care of automatic scaling in boost filters. If an insert I/O in fixed-point representation follows a boost filter in floating-point representation then the same scaling as in fixed-point arithmetic has to be done.

Dynamic Range Control. Dynamic range control is performed by a simple multiplicative weighting of the input signal with a control factor. The latter follows from calculating the peak and RMS value (root mean square) of the input signal. The number representation of the signal has no influence on the properties of the algorithm. Owing to the normalized mantissa in floating-point representation, some simplifications are produced while determining the control factor.

Mixing/Summation. While mixing signals into a stereo image, only multiplications and additions occur. Under the assumption of incoherent signals, an overload reserve can be estimated. This implies a reserve of 20/30 dB for 48/96 sources. For fixed-point representation the overload reserve is provided by a number of overflow bits in the accumulator of a DSP (*D*igital *S*ignal *P*rocessor). The properties of automatic scaling in floating-point arithmetic provide for overload reserves. For both number representations, the summation signal must be matched with the number representation of the output. While dealing with AES/EBU outputs or MADI outputs, both number representations are adjusted to fixed-point format. Similarly, within heterogeneous system solutions, it is logical to make heterogeneous use of both number representations, though corresponding number representations have to be converted.

Since the signal-to-noise ratio in fixed-point representation depends on the input level, a conversion from fixed-point to floating-point representation does not lead to a change of signal-to-noise ratio, i.e. the conversion does not improve the signal-to-noise ratio. Further signal processing with floating-point or fixed-point arithmetic does not alter the signal-to-noise ratio as long as the algorithms are chosen and programmed accordingly. Reconversion from floating-point to fixed-point representation again leads to a level-dependent signal-to-noise ratio.

As a consequence, for two-channel DSP systems which operate with AES/EBU or with analog inputs and outputs, and which are used for equalization, dynamic range control, room simulation etc., the above-mentioned discussion holds. These conclusions are also valid for digital mixing consoles for which digital inputs from AD converters or from multitrack machines are represented in fixed-point format (AES/EBU or MADI). The number representation for inserts and auxiliaries is specific to a system. Digital AES/EBU (or MADI) inputs and outputs are realized in fixed-point number representation.

2.5   Java Applet – Quantization, Dither, and Noise Shaping

This applet shown in Fig. 2.44 demonstrates audio effects resulting from quantization. It is designed for a first insight into the perceptual effects of quantizing an audio signal.

The following functions can be selected on the lower right of the graphical user interface:

- Quantizer

    - word-length w leads to quantization step size $Q = 2^{w-1}$.

- Dither

    - rect dither – uniform probability density function
    - tri dither – triangular probability density function

- high-pass dither – triangular probability density function and high-pass power spectral density.

- Noise shaping

    - first-order $H(z) = z^{-1}$
    - second-order $H(z) = -2z^{-1} + z^{-2}$
    - psychoacoustic noise shaping.

You can choose between two predefined audio files from our web server (*audio1.wav* or *audio2.wav*) or your own local wav file to be processed [Gui05].

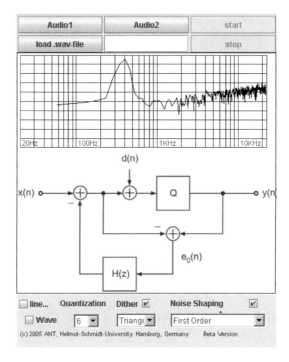

Figure 2.44 Java applet – quantization, dither, and noise shaping.

2.6 Exercises

1. Quantization

1. Consider a 100 Hz sine wave $x(n)$ sampled with $f_S = 44.1$ kHz, $N = 1024$ samples and $w = 3$ bits (word-length). What is the number of quantization levels? What is the quantization step Q when the signal is normalized to $-1 \leq x(n) < 1$. Show graphically how quantization is performed. What is the maximum error for this 3-bit quantizer? Write Matlab code for quantization with rounding and truncation.

2. Derive the mean value, the variance and the peak factor P_F of sequence $e(n)$, if the signal has a uniform probability density function in the range $-Q/2 < e(n) < -Q/2$. Derive the signal-to-noise ratio for this case. What will happen if we increase our word-length by one bit?

3. As the input signal level decreases from maximum amplitude to very low amplitudes, the error signal becomes more audible. Describe the error calculated above when w decreases to 1 bit? Is the classical quantization model still valid? What can be done to avoid this distortion?

4. Write Matlab code for a quantizer with $w = 16$ bits with rounding and truncation.

   - Plot the nonlinear transfer characteristic and the error signal when the input signal covers the range $3Q < x(n) < 3Q$.
   - Consider the sine wave $x(n) = A \sin(2\pi (f/f_S)n)$, $n = 0, \ldots, N-1$, with $A = Q$, $f/f_S = 64/N$ and $N = 1024$. Plot the output signal $(n = 0, \ldots, 99)$ of a quantizer with rounding and truncation in the time domain and the frequency domain.
   - Compute for both quantization types the quantization error and the signal-to-noise ratio.

2. Dither

1. What is dither and when do we have to use it?

2. How do we perform dither and what kinds of dither are there?

3. How do we obtain a triangular high-pass dither and why do we prefer it to other dithers?

4. Matlab: Generate corresponding dither signals for rectangular, triangular and triangular high-pass.

5. Plot the amplitude distribution and the spectrum of the output $x_Q(n)$ of a quantizer for every dither type.

3. Noise Shaping

1. What is noise shaping and when do we do it?

2. Why is it necessary to dither during noise shaping and how do we do this?

3. Matlab: The first noise shaper used is without dither and assumes that the transfer function in the feedback structure can be first-order $H(z) = z^{-1}$ or second-order $H(z) = -2z^{-1} + z^{-2}$. Plot the output $x_Q(n)$ and the error signal $e(n)$ and its spectrum. Show with a plot the shape of the error signal.

4. The same noise shaper is now used with a dither signal. Is it really necessary to dither with noise shaping? Where would you add your dither in the flow graph to achieve better results?

5. In the feedback structure we now use a psychoacoustic-based noise shaper which uses the Wannamaker filter coefficients

$$h_3 = [1.623, -0.982, 0.109],$$
$$h_5 = [2.033, -2.165, 1.959, -1.590, 0.6149],$$
$$h_9 = [2.412, -3.370, 3.937, -4.174, 3.353, -2.205, 1.281, -0.569, 0.0847].$$

Show with a Matlab plot the shape of the error with this filter.

References

[DeK03]   D. De Koning, W. Verhelst: *On Psychoacoustic Noise Shaping for Audio Requantization*, Proc. ICASSP-03, Vol. 5, pp. 453–456, April 2003.

[Ger89]   M. A. Gerzon, P. G. Craven: *Optimal Noise Shaping and Dither of Digital Signals*, Proc. 87th AES Convention, Preprint No. 2822, New York, October 1989.

[Gui05]   M. Guillemard, C. Ruwwe, U. Zölzer: *J-DAFx – Digital Audio Effects in Java*, Proc. 8th Int. Conference on Digital Audio Effects (DAFx-05), pp. 161–166, Madrid, 2005.

[Hel07]   C. R. Helmrich, M. Holters, U. Zölzer: *Improved Psychoacoustic Noise Shaping for Requantization of High-Resolution Digital Audio*, AES 31st International Conference, London, June 2007.

[ISO389]   ISO 389-7:2005: *Acoustics – Reference Zero for the Calibration of Audiometric Equipment – Part 7: Reference Threshold of Hearing Under Free-Field and Diffuse-Field Listening Conditions*, Geneva, 2005.

[Lip86]   S. P. Lipshitz, J. Vanderkoy: *Digital Dither*, Proc. 81st AES Convention, Preprint No. 2412, Los Angeles, November 1986.

[Lip92]   S. P. Lipshitz, R. A. Wannamaker, J. Vanderkoy: *Quantization and Dither: A Theoretical Survey*, J. Audio Eng. Soc., Vol. 40, No. 5, pp. 355–375, May 1992.

[Lip93]   S. P. Lipshitz, R. A. Wannamaker, J. Vanderkooy: *Dithered Noise Shapers and Recursive Digital Filters*, Proc 94th AES Convention, Preprint No. 3515, Berlin, March 1993.

[Sha48]   C. E. Shannon: *A Mathematical Theory of Communication*, Bell Systems, Techn. J., pp. 379–423, 623–656, 1948.

[Sri77]   A. B. Sripad, D. L. Snyder: *A Necessary and Sufficient Condition for Quantization Errors to be Uniform and White*, IEEE Trans. ASSP, Vol. 25, pp. 442–448, October 1977.

[Ter79]   E. Terhardt: *Calculating Virtual Pitch*, Hearing Res., Vol. 1, pp. 155–182, 1979.

[Van89]   J. Vanderkooy, S. P. Lipshitz: *Digital Dither: Signal Processing with Resolution Far Below the Least Significant Bit*, Proc. AES Int. Conf. on Audio in Digital Times, pp. 87–96, May 1989.

[Wan92]   R. A. Wannamaker: *Psychoacoustically Optimal Noise Shaping*, J. Audio Eng. Soc., Vol. 40, No. 7/8, pp. 611–620, July/August 1992.

[Wan00]   R. A. Wannamaker, S. P. Lipshitz, J. Vanderkooy, J. N. Wright: *A Theory of Nonsubtractive Dither*, IEEE Trans. Signal Processing, Vol. 48, No. 2, pp. 499–516, 2000.

[Wid61]   B. Widrow: *Statistical Analysis of Amplitude-Quantized Sampled-Data Systems*, Trans. AIEE, Pt. II, Vol. 79, pp. 555–568, January 1961.

Chapter 3

AD/DA Conversion

The conversion of a continuous-time function $x(t)$ (voltage, current) into a sequence of numbers $x(n)$ is called analog-to-digital conversion (AD conversion). The reverse process is known as digital-to-analog conversion (DA conversion). The time-sampling of a function $x(t)$ is described by *Shannon's sampling theorem*. This states that a continuous-time signal with bandwidth f_B can be sampled with a sampling rate $f_S > 2f_B$ without changing the information content in the signal. The original analog signal is reconstructed by low-pass filtering with bandwidth f_B. Besides time-sampling, the nonlinear procedure of digitizing the continuous-valued amplitude (quantization) of the sampled signal occurs. In Section 3.1 basic concepts of Nyquist sampling, oversampling and delta-sigma modulation are presented. In Sections 3.2 and 3.3 principles of AD and DA converter circuits are discussed.

3.1 Methods

3.1.1 Nyquist Sampling

The sampling of a signal with sampling rate $f_S > 2f_B$ is called Nyquist sampling. The schematic diagram in Fig. 3.1 shows the procedure. The band-limiting of the input at $f_S/2$ is carried out by an analog low-pass filter (Fig. 3.1a). The following sample-and-hold circuit samples the band-limited input at a sampling rate f_S. The constant amplitude of the time function over the sampling period $T_S = 1/f_S$ is converted to a number sequence $x(n)$ by a quantizer (Fig. 3.1b). This number sequence is fed to a digital signal processor (DSP) which performs signal processing algorithms. The output sequence $y(n)$ is delivered to a DA converter which gives a staircase as its output (Fig. 3.1c). Following this, a low-pass filter gives the analog output $y(t)$ (Fig. 3.1d). Figure 3.2 demonstrates each step of AD/DA conversion in the frequency domain. The individual spectra in Fig. 3.2a–d correspond to the outputs in Fig. 3.1a–d.

After band-limiting (Fig. 3.2a) and sampling, a periodic spectrum with period f_S of the sampled signal is obtained as shown in Fig. 3.2b. Assuming that consecutive quantization errors $e(n)$ are statistically independent of each other, the noise power has

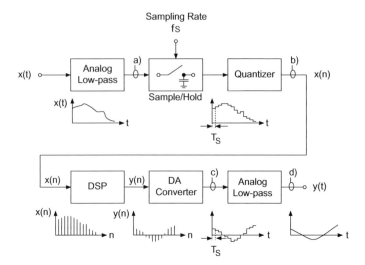

Figure 3.1 Schematic diagram of Nyquist sampling.

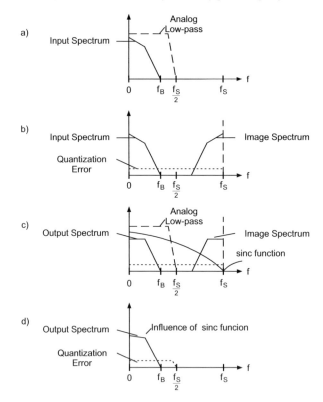

Figure 3.2 Nyquist sampling – interpretation in the frequency domain.

a spectral uniform distribution in the frequency domain $0 \le f \le f_S$. The output of the DA converter still has a periodic spectrum. However, this is weighted with the sinc function (sinc $= \sin(x)/x$), of the sample-and-hold circuit (Fig. 3.2c). The zeros of the sinc function are at multiples of the sampling rate f_S. In order to reconstruct the output (Fig. 3.2d), the image spectra are eliminated by an analog low-pass of sufficient stop-band attenuation (see Fig. 3.2c).

The problems of Nyquist sampling lie in the steep band-limiting filter characteristics (anti-aliasing filter) of the analog input filter and the analog reconstruction filter (anti-imaging filter) of similar filter characteristics and sufficient stop-band attenuation. Further, sinc distortion due to the sample-and-hold circuit needs to be compensated for.

3.1.2 Oversampling

In order to increase the resolution of the conversion process and reduce the complexity of analog filters, oversampling techniques are employed. Owing to the spectral uniform distribution of quantization error between 0 and f_S (see Fig. 3.3a), it is possible to reduce the power spectral density in the pass-band $0 \le f \le f_B$ through oversampling by a factor L, i.e. with the new sampling rate Lf_S (see Fig. 3.3b). For identical quantization step size Q, the shaded areas (quantization error power σ_E^2) in Fig. 3.3a and Fig. 3.3b are equal. The increase in the signal-to-noise ratio can also be observed in Fig. 3.3.

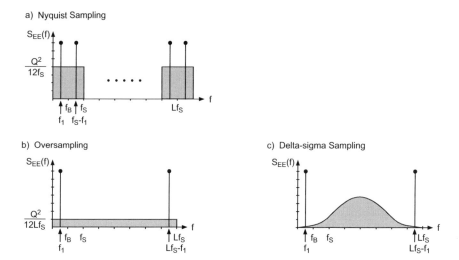

Figure 3.3 Influence of oversampling and delta-sigma technique on power spectral density of quantization error and on input sinusoid with frequency f_1.

It follows that in the pass-band at a sampling rate of $f_S = 2f_B$ the power spectral density given by

$$S_{EE}(f) = \frac{Q^2}{12 f_S} \qquad (3.1)$$

leads to the noise power

$$N_B^2 = \sigma_E^2 = 2 \int_0^{f_B} S_{EE}(f)\, df = \frac{Q^2}{12}. \tag{3.2}$$

Owing to oversampling by a factor of L, a reduction of the power spectral density given by

$$S_{EE}(f) = \frac{Q^2}{12Lf_S} \tag{3.3}$$

is obtained (see Fig. 3.3b). With $f_S = 2f_B$, the error power in the audio band is given by

$$N_B^2 = 2f_B \frac{Q^2}{12Lf_S} = \frac{Q^2}{12}\frac{1}{L}. \tag{3.4}$$

The signal-to-noise ratio (with $P_F = \sqrt{3}$) owing to oversampling can now be expressed as

$$\text{SNR} = 6.02 \cdot w + 10\log_{10}(L)\ \text{dB}. \tag{3.5}$$

Figure 3.4a shows a schematic diagram of anoversampling AD converter. Owing to oversampling, the analog band-limiting low-pass filter can have a wider transition bandwidth as shown in Fig. 3.4b. The quantization error power is distributed between 0 and the sampling rate Lf_S. To reduce the sampling rate, it is necessary to limit the bandwidth with a digital low-pass filter (see Fig. 3.4c). After this, the samplingrate is reduced by a factor L (see Fig. 3.4d) by taking every Lth output sample of the digital low-pass filter [Cro83, Vai93, Fli00].

Figure 3.5a shows a schematic diagram of an oversampling DA converter. The sampling rate is first increased by a factor of L. For this purpose, $L-1$ zeros are introduced between two consecutive input values [Cro83, Vai93, Fli00]. The following digital filter eliminates all image spectra (Fig. 3.5b) except the base-band spectrum and spectra at multiples of Lf_S (Fig. 3.5c). It interpolates $L-1$ samples between two input samples. The w-bit DA converter operates at a sampling rate Lf_S. Its output is fed to an analog reconstruction filter which eliminates the image spectra at multiples of Lf_S.

3.1.3 Delta-sigma Modulation

Delta-sigma modulation using oversampling is a conversion strategy derived from delta modulation. In delta modulation (Fig. 3.6a), the difference between the input $x(t)$ and signal $x_1(t)$ is converted into a 1-bit signal $y(n)$ at a very high sampling rate Lf_S. The sampling rate is higher than the necessary Nyquist rate f_S. The quantized signal $y(n)$ gives the signal $x_1(t)$ via an analog integrator. The demodulator consists of an integrator and a reconstruction low-pass filter.

The extension to delta-sigma modulation [Ino63] involves shifting the integrator from the demodulator to the input of the modulator (see Fig. 3.6b). With this, it is possible to combine the two integrators as a single integrator after addition (see Fig. 3.7a). The corresponding signals are shown in Fig. 3.8.

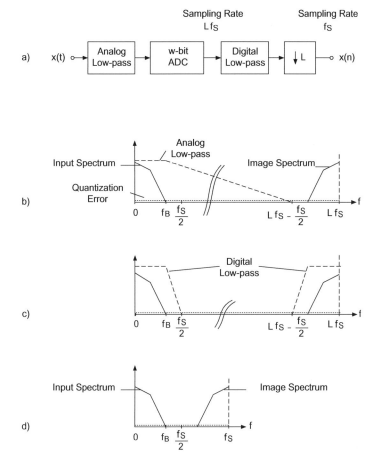

Figure 3.4 Oversampling AD converter and sampling rate reduction.

A time-discrete model of the delta-sigma modulator is given in Fig. 3.7b. The Z-transform of the output signal $y(n)$ is given by

$$Y(z) = \frac{H(z)}{1 + H(z)} X(z) + \frac{1}{1 + H(z)} E(z) \approx X(z) + \frac{1}{1 + H(z)} E(z). \qquad (3.6)$$

For a large gain factor of the system $H(z)$, the input signal will not be affected. In contrast, the quantization error is shaped by the filter term $1/[(1 + H(z)]$.

Schematic diagrams of delta-sigma AD/DA conversion are shown in Figs 3.9 and 3.10. For delta-sigma AD converters, a digital low-pass filter and a downsampler with factor L are used to reduce the sampling rate Lf_S to f_S. The 1-bit input to the digital low-pass filter leads to a w-bit output $x(n)$ at a sampling rate f_S. The delta-sigma DA converter consists of an upsampler with factor L, a digital low-pass filter to eliminate the mirror spectra and a delta-sigma modulator followed by an analog reconstruction low-pass filter. In order to illustrate noise shaping in delta-sigma modulation in detail, first- and second-order systems as well as multistage techniques are investigated in the following sections.

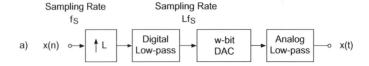

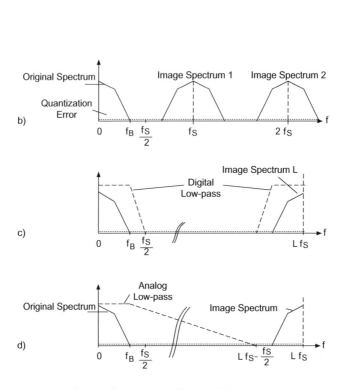

Figure 3.5 Oversampling and DA conversion.

First-order Delta-sigma Modulator

A time-discrete model of a first-order delta-sigma modulator is shown in Fig. 3.11.
    The difference equation for the output $y(n)$ is given by

$$y(n) = x(n-1) + e(n) - e(n-1). \tag{3.7}$$

The corresponding Z-transform leads to

$$Y(z) = z^{-1}X(z) + E(z) \underbrace{(1 - z^{-1})}_{H_E(z)}. \tag{3.8}$$

The power density spectrum of the error signal $e_1(n) = e(n) - e(n-1)$ is

$$S_{E_1 E_1}(e^{j\Omega}) = S_{EE}(e^{j\Omega})|1 - e^{-j\Omega}|^2 = S_{EE}(e^{j\Omega}) 4 \sin^2\left(\frac{\Omega}{2}\right), \tag{3.9}$$

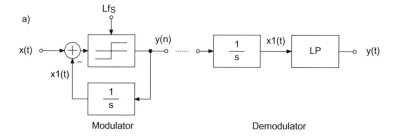

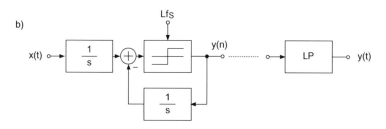

Figure 3.6 Delta modulation and displacement of integrator.

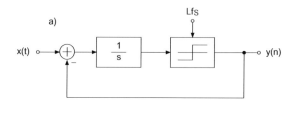

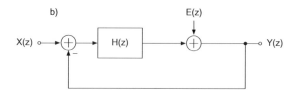

Figure 3.7 Delta-sigma modulation and time-discrete model.

where $S_{EE}(e^{j\Omega})$ denotes the power density spectrum of the quantization error $e(n)$. The error power in the frequency band $[-f_B, f_B]$, with $S_{EE}(f) = Q^2/12Lf_S$, can be written as

$$N_B^2 = S_{EE}(f)2 \int_0^{f_B} 4 \sin^2\left(\pi \frac{f}{Lf_S}\right) df \tag{3.10}$$

$$\simeq \frac{Q^2}{12} \frac{\pi^2}{3} \left(\frac{2f_B}{Lf_S}\right)^3. \tag{3.11}$$

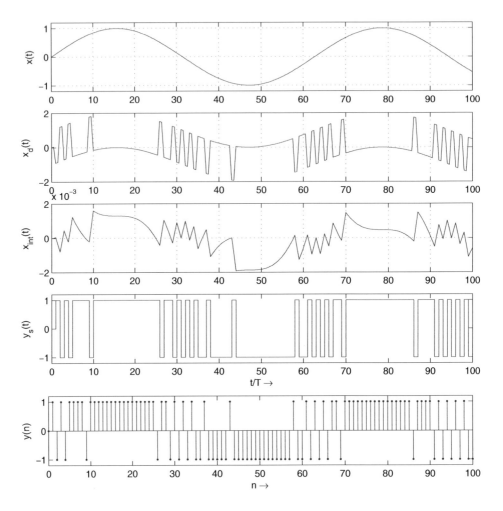

Figure 3.8 Signals in delta-sigma modulation.

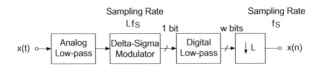

Figure 3.9 Oversampling delta-sigma AD converter.

With $f_S = 2 f_B$, we get

$$N_B^2 = \frac{Q^2}{12} \frac{\pi^2}{3} \left(\frac{1}{L} \right)^3. \tag{3.12}$$

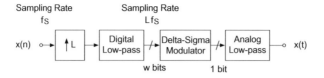

Figure 3.10 Oversampling delta-sigma DA converter.

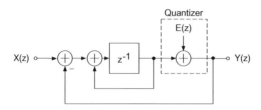

Figure 3.11 Time-discrete model of a first-order delta-sigma modulator.

Second-order Delta-sigma Modulator

For the second-order delta-sigma modulator [Can85], shown in Fig. 3.12, the difference equation is expressed as

$$y(n) = x(n-1) + e(n) - 2e(n-1) + e(n-2) \tag{3.13}$$

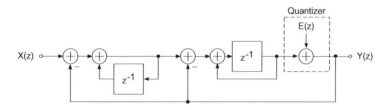

Figure 3.12 Time-discrete model of a second-order delta-sigma modulator.

and the Z-transform is given by

$$Y(z) = z^{-1}X(z) + E(z) \underbrace{(1 - 2z^{-1} + z^{-2})}_{H_E(z) = (1 - z^{-1})^2}. \tag{3.14}$$

The power density spectrum of the error signal $e_1(n) = e(n) - 2e(n-1) + e(n-2)$ can be written as

$$\begin{aligned}
S_{E_1 E_1}(e^{j\Omega}) &= S_{EE}(e^{j\Omega})|1 - e^{-j\Omega}|^4 \\
&= S_{EE}(e^{j\Omega})\left[4\sin^2\left(\frac{\Omega}{2}\right)\right]^2 \\
&= S_{EE}(e^{j\Omega})4[1 - \cos(\Omega)]^2.
\end{aligned} \tag{3.15}$$

The error power in the frequency band $[-f_B, f_B]$ is given by

$$N_B^2 = S_{EE}(f)2 \int_0^{f_B} 4[1 - \cos(\Omega)]^2 \, df \qquad (3.16)$$

$$\simeq \frac{Q^2}{12} \frac{\pi^4}{5} \left(\frac{2f_B}{Lf_S}\right)^5, \qquad (3.17)$$

and with $f_S = 2f_B$ we obtain

$$N_B^2 = \frac{Q^2}{12} \frac{\pi^4}{5} \left(\frac{1}{L}\right)^5. \qquad (3.18)$$

Multistage Delta-sigma Modulator

A multistage delta-sigma modulator (MASH, [Mat87]) is shown in Fig. 3.13.

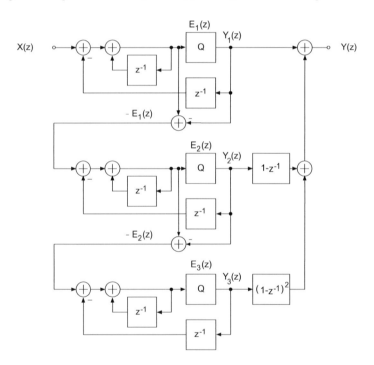

Figure 3.13 Time-discrete model of a multistage delta-sigma modulator.

The Z-transforms of the output signals $y_i(n)$, $i = 1, 2, 3$, are given by

$$Y_1(z) = X(z) + (1 - z^{-1})E_1(z), \qquad (3.19)$$

$$Y_2(z) = -E_1(z) + (1 - z^{-1})E_2(z), \qquad (3.20)$$

$$Y_3(z) = -E_2(z) + (1 - z^{-1})E_3(z). \qquad (3.21)$$

The Z-transform of the output obtained by addition and filtering leads to

$$
\begin{aligned}
Y(z) &= Y_1(z) + (1 - z^{-1})Y_2(z) + (1 - z^{-1})^2 Y_3(z) \\
&= X(z) + (1 - z^{-1})E_1(z) - (1 - z^{-1})E_1(z) \\
&\quad + (1 - z^{-1})^2 E_2(z) - (1 - z^{-1})^2 E_2(z) + (1 - z^{-1})^3 E_3(z) \\
&= X(z) + \underbrace{(1 - z^{-1})^3}_{H_E(z)} E_3(z).
\end{aligned}
\tag{3.22}
$$

The error power in the frequency band $[-f_B,\, f_B]$,

$$
N_B^2 = \frac{Q^2}{12}\frac{\pi^6}{7}\left(\frac{2f_B}{Lf_S}\right)^7,
\tag{3.23}
$$

with $f_S = 2f_B$, gives the total noise power

$$
N_B^2 = \frac{Q^2}{12}\frac{\pi^6}{7}\left(\frac{1}{L}\right)^7.
\tag{3.24}
$$

The error transfer functions in Fig. 3.14 show the noise shaping for three types of delta-sigma modulations as discussed before. The error power is shifted toward higher frequencies.

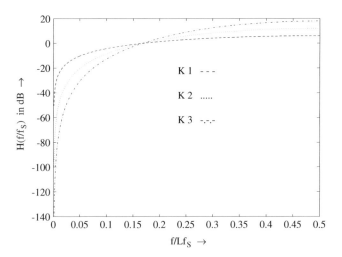

Figure 3.14 $H_E(z) = (1 - z^{-1})^K$ with $K = 1, 2, 3$.

The improvement of signal-to-noise ratio by pure oversampling and delta-sigma modulation (first, second and third order) is shown in Fig. 3.15. For the general case of a kth-order delta-sigma conversion with oversampling factor L one can derive the signal-to-noise ratio as

$$
\text{SNR} = 6.02 \cdot w - 10\log_{10}\left(\frac{\pi^{2k}}{2k+1}\right) + (2k+1)10\log_{10}(L)\ \text{dB}.
\tag{3.25}
$$

Here w denotes the quantizer word-length of the delta-sigma modulator. The signal quantization after digital low-pass filtering and downsampling by L can be performed with (3.25) according to the relation $w = \mathrm{SNR}/6$.

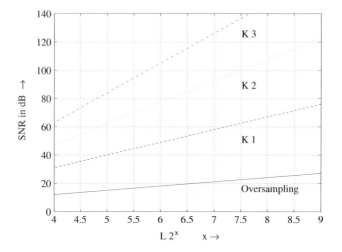

Figure 3.15 Improvement of signal-to-noise ratio as a function of oversampling and noise shaping ($L = 2^x$).

Higher-order Delta-sigma Modulator

A widening of the stop-band for the high-pass transfer function of the quantization error is achieved with higher-order delta-sigma modulation [Cha90]. Besides the zeros at $z = 1$, additional zeros are placed on the unit circle. Also, poles are integrated into the transfer function. A time-discrete model of a higher-order delta-sigma modulator is shown in Fig. 3.16.

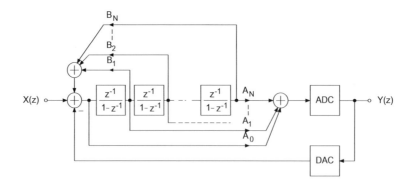

Figure 3.16 Higher-order delta-sigma modulator.

The transfer function in Fig. 3.16 can be written as

$$
\begin{aligned}
H(z) &= \frac{A_0 + A_1 \frac{z^{-1}}{1-z^{-1}} + A_2 \left(\frac{z^{-1}}{1-z^{-1}}\right)^2 + \cdots}{1 - B_1 \frac{z^{-1}}{1-z^{-1}} - B_2 \left(\frac{z^{-1}}{1-z^{-1}}\right)^2 + \cdots} \\
&= \frac{A_0(z-1)^N + A_1(z-1)^{N-1} + \cdots + A_N}{(z-1)^N - B_1(z-1)^{N-1} - \cdots - B_N} \\
&= \frac{\sum_{i=0}^{N} A_i (z-1)^{N-i}}{(z-1)^N - \sum_{i=1}^{N} B_i (z-1)^{N-i}}.
\end{aligned} \tag{3.26}
$$

The Z-transform of the output is given by

$$
Y(z) = \frac{H(z)}{1 + H(z)} X(z) + \frac{1}{1 + H(z)} E(z) \tag{3.27}
$$
$$
= H_X(z) X(z) + H_E(Z) E(z). \tag{3.28}
$$

The transfer function for the input is

$$
H_X(z) = \frac{\sum_{i=0}^{N} A_i (z-1)^{N-i}}{(z-1)^N - \sum_{i=1}^{N} B_i (z-1)^{N-i} + \sum_{i=0}^{N} A_i (z-1)^{N-i}}, \tag{3.29}
$$

and the transfer function for the error signal is given by

$$
H_E(z) = \frac{(z-1)^N - \sum_{i=1}^{N} B_i (z-1)^{N-i}}{(z-1)^N - \sum_{i=1}^{N} B_i (z-1)^{N-i} + \sum_{i=0}^{N} A_i (z-1)^{N-i}}. \tag{3.30}
$$

For Butterworth or Chebyshev filter designs, the frequency responses as shown in Fig. 3.17 are obtained for the error transfer functions. As a comparison, the frequency responses of first-, second- and third-order delta-sigma modulation are shown. The widening of the stop-band for Butterworth and Chebyshev filters can be observed from Fig. 3.18.

Decimation Filter

Decimation filters for AD conversion and interpolation filters for DA conversion are implemented with multirate systems [Fli00]. The necessary downsampler and upsampler are simple systems. For the former, every nth sample is taken out of the input sequence. For the latter, $n - 1$ zeros are inserted between two input samples. For decimation, band-limiting is performed by $H(z)$ followed by sampling rate reduction by a factor L. This procedure can be implemented in stages (see Fig. 3.19). The use of easy-to-implement filter structures at high sampling rates, like comb filters with transfer function

$$
H_1(z) = \frac{1}{L} \frac{1 - z^{-L}}{1 - z^{-1}} \tag{3.31}
$$

(shown in Fig. 3.20), allows simple implementation needing only delay systems and additions. In order to increase the stop-band attenuation, a series of comb filters is used so

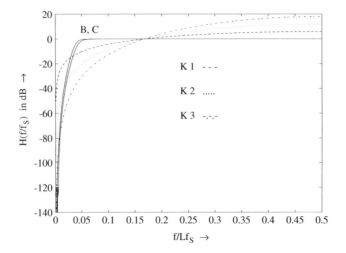

Figure 3.17 Comparison of different transfer functions of error signal.

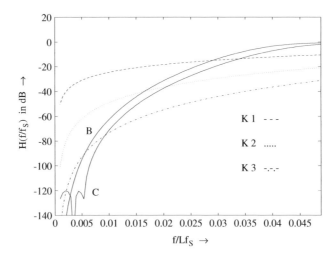

Figure 3.18 Transfer function of the error signal in stop-band.

that

$$H_1^M(z) = \left[\frac{1}{L} \frac{1 - z^{-L}}{1 - z^{-1}} \right]^M \tag{3.32}$$

is obtained.

Besides additions at high sampling rates, complexity can be reduced further. Owing to sampling rate reduction by a factor of L, the numerator $(1 - z^{-L})$ can be moved so that it is placed after the downsampler (see Fig. 3.21). For a series of comb filters, the

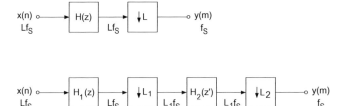

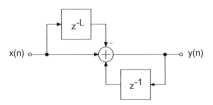

Figure 3.19 Several stages of sampling rate reduction.

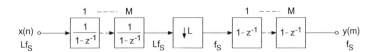

Figure 3.20 Signal flow diagram of a comb filter.

structure in Fig. 3.22 results. M simple recursive accumulators have to be performed at the high sampling rate Lf_S. After this, downsampling by a factor L is carried out. The M nonrecursive systems are calculated with the output sampling rate f_S.

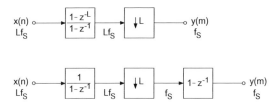

Figure 3.21 Comb filter for sampling rate reduction.

Figure 3.22 Series of comb filters for sampling rate reduction.

Figure 3.23a shows the frequency responses of a series of comb filters ($L = 16$). Figure 3.23b shows the resulting frequency response for the quantization error of a third-order delta-sigma modulator connected in series with a comb filter $H_1^4(z)$. The system delay owing to filtering and sampling rate reduction is given by

$$t_D = \frac{N-1}{2} \cdot \frac{1}{Lf_S}. \tag{3.33}$$

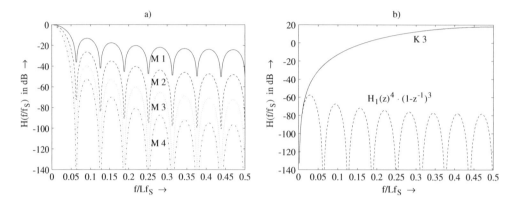

Figure 3.23 (a) Transfer function $H_1^M(z) = \left[\frac{1}{16}\frac{1-z^{-16}}{1-z^{-1}}\right]^M$ with $M = 1\ldots4$. (b) Third-order delta-sigma modulation and in series with $H_1^4(z)$.

Example: Delay time of conversion process (latency time)

1. Nyquist conversion

$$f_S = 48\text{ kHz}$$
$$t_D = \frac{1}{f_S} = 20.83\text{ μs}.$$

2. Delta-sigma modulation with single-stage downsampling

$$L = 64$$
$$f_S = 48\text{ kHz}$$
$$N = 4096$$
$$t_D = 665\text{ μs}.$$

3. Delta-sigma modulation with two-stage downsampling

$$L = 64$$
$$f_S = 48\text{ kHz}$$
$$L_1 = 16$$
$$L_2 = 4$$
$$N_1 = 61$$
$$N_2 = 255$$
$$t_{D_1} = 9.76\text{ μs}$$
$$t_{D_2} = 662\text{ μs}.$$

3.2 AD Converters

The choice of an AD converter for a certain application is influenced by a number of factors. It mainly depends on the necessary resolution for a given conversion time. Both of these depend upon each other and are decisively influenced by the architecture of the AD converter. For this reason, the specifications of an AD converter are first discussed. This is followed by circuit principles which influence the mutual dependence of resolution and conversion time.

3.2.1 Specifications

In the following, the most important specifications for AD conversion are presented.

Resolution. The resolution for a given word-length w of an AD converter determines the smallest amplitude

$$x_{min} = Q = x_{max} \, 2^{-(w-1)}, \tag{3.34}$$

which is equal to the quantization step Q.

Conversion Time. The minimum sampling period $T_S = 1/f_S$ between two samples is called the conversion time.

Sample-and-hold Circuit. Before quantization, the time-continuous function is sampled with the help of a sample-and-hold circuit, as shown in Fig. 3.24a.

The sampling period T_S is divided into the sampling time t_S in which the output voltage U_2 follows the input voltage U_1, and the hold time t_H. During the hold time the output voltage U_2 is constant and is converted into a binary word by quantization.

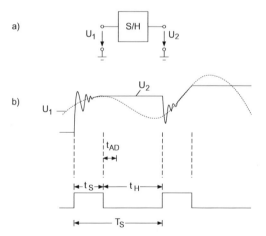

Figure 3.24 (a) Sample-and-hold circuit. (b) Input and output with clock signal. (t_S = sampling time, t_H = hold time, t_{AD} = aperture delay.)

Aperture Delay. The time t_{AD} elapsed between start of hold and actual hold mode (see Fig. 3.24b) is called the aperture delay.

Aperture Jitter. The variation in aperture delay from sample to sample is called the aperture jitter t_{ADJ}. The influence of aperture jitter limits the useful bandwidth of the sampled signal. This is because at high frequency a deterioration of the signal-to-noise ratio occurs. Assuming a Gaussian PDF aperture jitter, the signal-to-noise ratio owing to aperture jitter as a function of frequency f can be written as

$$\text{SNR}_J = -20 \log_{10}(2\pi f t_{ADJ}) \text{ dB.} \tag{3.35}$$

Offset Error and Gain Error. The offset and gain errors of an AD converter are shown in Fig. 3.25. The offset error results in a horizontal displacement of the real curve compared with the dashed ideal curve of an AD converter. The gain error is expressed as the deviation from the ideal gradient of the curve.

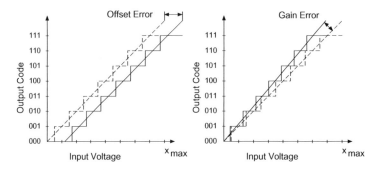

Figure 3.25 Offset error and gain error.

Differential Nonlinearity. The differential nonlinearity

$$\text{DNL} = \frac{\Delta x / Q}{\Delta x_Q} - 1 \text{ LSB} \tag{3.36}$$

describes the error of the step size of a certain code word in LSB units. For ideal quantization, the increase Δx in the input voltage up to the next output code x_Q is equal to the quantization step Q (see Fig. 3.26). The difference of two consecutive output codes is denoted by Δx_Q. When the output code changes from 010 to 011, the step size is 1.5 LSB and therefore the differential nonlinearity $\text{DNL} = 0.5$ LSB. The step size between the codes 011 and 101 is 0 LSB and the code 200 is missing. The differential nonlinearity is $\text{DNL} = -1$ LSB.

Integral Nonlinearity. The integral nonlinearity (INL) describes the error between the quantized and the ideal continuous value. This error is given in LSB units. It arises owing to the accumulated error of the step size. This (see Fig. 3.27) changes itself continuously from one output code to another.

Monotonicity. The progressive increase in quantizer output code for a continuously increasing input voltage and progressive decrease in quantizer output code for a continuously decreasing input voltage is called monotonicity. An example of non-monotonic behavior is shown in Fig. 3.28 where one output code does not occur.

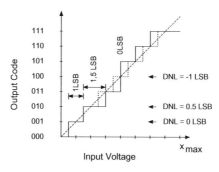

Figure 3.26 Differential nonlinearity.

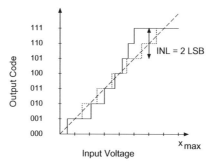

Figure 3.27 Integral nonlinearity.

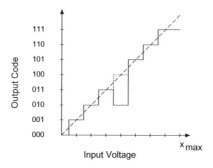

Figure 3.28 Monotonicity.

Total Harmonic Distortion. The harmonic distortion is calculated for an AD converter at full range with a sinusoid ($X_1 = 0$ dB) of given frequency. The selective measurement of harmonics of the second to the ninth order are used to compute

$$\text{THD} = 20 \log \sqrt{\sum_{n=2}^{\infty} [10^{(-X_n/20)}]^2} \ \text{dB} \tag{3.37}$$

$$= \sqrt{\sum_{n=2}^{\infty} [10^{(-X_n/20)}]^2} \cdot 100\%, \tag{3.38}$$

where X_n are the harmonics in dB.

THD+N: Total Harmonic Distortion plus Noise. For the calculation of harmonic distortion plus noise, the test signal is suppressed by a stop-band filter. The measurement of harmonic distortion plus noise is performed by measuring the remaining broad-band noise signal which consists of integral and differential nonlinearity, missing codes, aperture jitter, analog noise and quantization error.

3.2.2 Parallel Converter

Parallel Converter. A direct method for AD conversion is called parallel conversion (flash converter). In parallel converters, the output voltage of the sample-and-hold circuit is compared with a reference voltage U_R with the help of $2^w - 1$ comparators (see Fig. 3.29). The sample-and-hold circuit is controlled with sampling rate f_S so that, during the hold time t_H, a constant voltage at the output of the sample-and-hold circuit is available. The outputs of the comparators are fed at sampling clock rate into a $(2^w - 1)$-bit register and converted by a coding logic to a w-bit data word. This is fed at sampling clock rate to an output register. The sampling rates that can be achieved lie between 1 and 500 MHz for a resolution of up to 10 bits. Owing to the large number of comparators, the technique is not feasible for high precision.

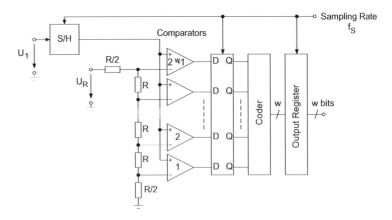

Figure 3.29 Parallel converter.

Half-flash Converter. In half-flash AD converters (Fig. 3.30), two m-bit parallel converters are used in order to convert two different ranges. The first m-bit AD converter gives a digital output word which is converted into an analog voltage using an m-bit DA converter. This voltage is now subtracted from the output voltage of the sample-and-hold circuit. The difference voltage is digitized with a second m-bit AD converter. The rough and fine quantization leads to a w-bit data word with a subsequent logic.

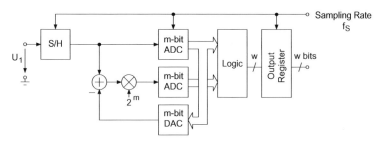

Figure 3.30 Half-flash AD converter.

Subranging Converter. A combination of direct conversion and sequential procedure is carried out for subranging AD converters (see Fig. 3.31). In contrast to the half-flash converter, only one parallel converter is required. The switches S_1 and S_2 take the values of 0 and 1. First the output voltage of a sample-and-hold circuit and then the difference voltage amplified by a factor 2^m is fed to an m-bit AD converter. The difference voltage is formed with the help of the output voltage of an m-bit DA converter and the output voltage of the sample-and-hold circuit. The conversion rates lie between 100 kHz and 40 MHz where a resolution of up to 16 bits is achieved.

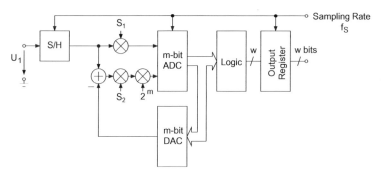

Figure 3.31 Subranging AD converter.

3.2.3 Successive Approximation

AD converters with successive approximation consist of the functional modules shown in Fig. 3.32. The analog voltage is converted into a w-bit word within w cycles. The converter consists of a comparator, a w-bit DA converter and logic for controlling the successive approximation.

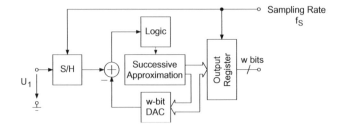

Figure 3.32 AD converter with successive approximation.

The conversion process is explained with the help of Fig. 3.33. First, it is checked whether a positive or negative voltage is present at the comparator. If it is positive, the output $+0.5U_R$ is fed to a DA converter to check whether the output voltage of the comparator is greater or less than $+0.5U_R$. Then, the output of $(+0.5 \pm 0.25)U_R$ is fed to the DA comparator. The output of the comparator is then evaluated. This procedure is performed w times and leads to a w-bit word.

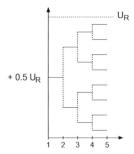

Figure 3.33 Successive approximation.

For a resolution of 12 bits, sampling rates of up to 1 MHz can be achieved. Higher resolutions of more than 16 bits are possible at a lower sampling rates.

3.2.4   Counter Methods

In contrast to the conversion techniques of the previous sections for high conversion rates, the following techniques are used for sampling rates smaller than 50 kHz.

Forward-backward Counter. A technique which operates like successive approximation is the forward-backward counter shown in Fig. 3.34. A logic controls a clocked forward-backward counter whose output data word provides an analog output voltage via a w-bit DA converter. The difference signal between this voltage and the output voltage of the sample-and-hold circuit determines the direction of counting. The counter stops when the corresponding output voltage of the DA converter is equal to the output voltage of the sample-and-hold circuit.

Single-slope Counter. The single-slope AD converter shown in Fig. 3.35 compares the output voltage of the sample-and-hold circuit with a voltage of a sawtooth generator. The

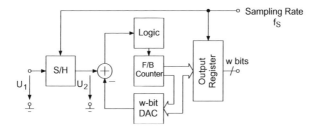

Figure 3.34 AD converter with forward-backward counter.

sawtooth generator is started every sampling period. As long as the input voltage is greater than the sawtooth voltage, the clock impulses are counted. The counter value corresponds to the digital value of the input voltage.

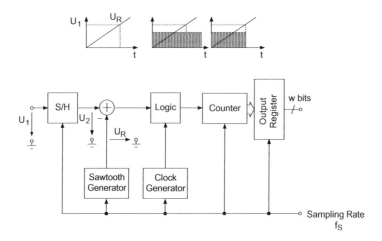

Figure 3.35 Single-slope AD converter.

Dual-slope Converter. A dual-slope AD converter is shown in Fig. 3.36. In the first phase in which a switch S_1 is closed for a counter period t_1, the output voltage of the sample-and-hold circuit is fed to an integrator of time-constant τ. During the second phase, the switch S_2 is closed and the switch S_1 is opened. The reference voltage is switched to the integrator and the time to reach a threshold is determined by counting the clock impulses by a counter. Figure 3.36 demonstrates this for three different voltages U_2. The slope during time t_1 is proportional to the output voltage U_2 of the sample-and-hold circuit, whereas the slope is constant when the reference voltage U_R is connected to the integrator. The ratio $U_2/U_R = t_2/t_1$ leads to the digital output word.

3.2.5 Delta-sigma AD Converter

The delta-sigma AD converter in Fig. 3.37 requires no sample-and-hold circuit owing to its high conversion rate. The analog band-limiting low-pass filter and the digital low-pass filter

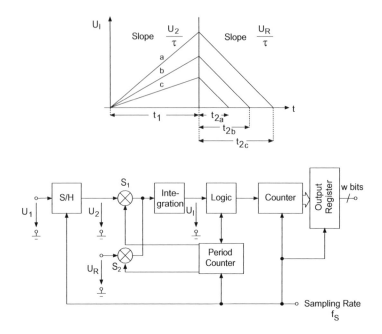

Figure 3.36 Dual-slope AD converter.

for downsampling to a sampling rate f_S are usually on the same circuit. The linear phase nonrecursive digital low-pass filter in Fig. 3.37 has a 1-bit input signal and leads to a w-bit output signal owing to the N filter coefficients $h_0, h_1, \ldots, h_{N-1}$ which are implemented with a word-length of w bits. The output signal of the filter results from the summation of the filter coefficients (0 or 1) of the nonrecursive low-pass filter. The downsampling by a factor L is performed by taking every Lth sample out of the filter and writing to the output register. In order to reduce the number of operations the filtering and downsampling can be performed only every Lth input sample.

Applications of delta-sigma AD converters are found at sampling rates of up to 100 kHz with a resolution of up to 24 bits.

3.3 DA Converters

Circuit principles for DA converters are mainly based on direct conversion techniques of the input code. Achievable sampling rates are accordingly high.

3.3.1 Specifications

The definitions of resolution, total harmonic distortion (THD) and total harmonic distortion plus noise (THD+N) correspond to those for AD converters. Further specifications are discussed in the following.

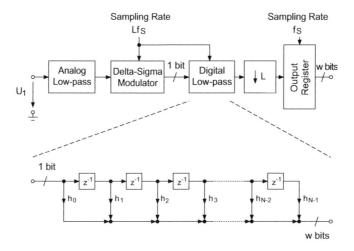

Figure 3.37 Delta-sigma AD converter.

Settling Time. The time interval between transferring a binary word and achieving the analog output value within a specific error range is called the settling time t_{SE}. The settling time determines the maximum conversion frequency $f_{S_{max}} = 1/t_{SE}$. Within this time, glitches between consecutive amplitude values can occur (see Fig. 3.38). With the help of a sample-and-hold circuit (deglitcher), the output voltage of the DA converter is sampled after the settling time and held.

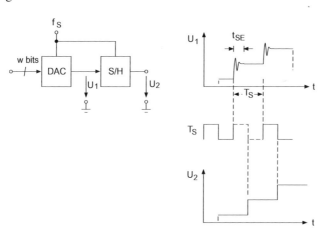

Figure 3.38 Settling time and sample-and-hold function.

Offset and Gain Error. The offset and gain errors of a DA converter are shown in Fig. 3.39.

Differential Nonlinearity. The differential nonlinearity for DA converters describes the step size error of a code word in LSB units. For ideal quantization, the increase Δx of the output voltage until the next code word corresponding to the output voltage is equal to

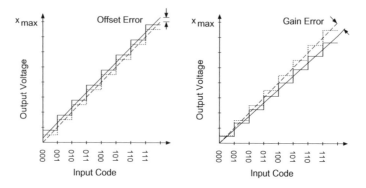

Figure 3.39 Offset and gain error.

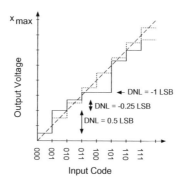

Figure 3.40 Differential nonlinearity.

the quantization step size Q (see Fig. 3.40). The difference between two consecutive input codes is termed Δx_Q. Differential nonlinearity is given by

$$\text{DNL} = \frac{\Delta x / Q}{\Delta x_Q} - 1 \text{ LSB}. \tag{3.39}$$

For the code steps from 001 to 010 as shown in Fig. 3.40, the step size is 1.5 LSB, and therefore the differential nonlinearity DNL $= 0.5$ LSB. The step size between the codes 010 and 100 is 0.75 LSB and DNL $= -0.25$. The step size for the code change from 011 to 100 is 0 LSB (DNL $= -1$ LSB).

Integral Nonlinearity. The integral nonlinearity describes the maximum deviation of the output voltage of a real DA converter from the ideal straight line (see Fig. 3.41).

Monotonicity. The continuous increase in the output voltage with increasing input code and the continuous decrease in the output voltage with decreasing input code is called monotonicity. A non-monotonic behavior is presented in Fig. 3.42.

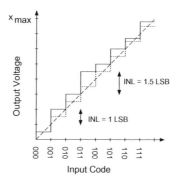

Figure 3.41 Integral nonlinearity.

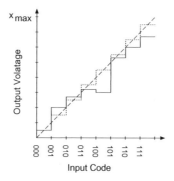

Figure 3.42 Monotonicity.

3.3.2   Switched Voltage and Current Sources

Switched Voltage Sources. The DA conversion with switched voltage sources shown in Fig. 3.43a is carried out with a reference voltage connected to a resistor network. The resistor network consists of 2^w resistors of equal resistance and is switched in stages to a binary-controlled decoder so that, at the output, a voltage U_2 is present corresponding to the input code. Figure 3.43b shows the decoder for a 3-bit input code 101.

Switched Current Sources. DA conversion with 2^w switched current sources is shown in Fig. 3.44. The decoder switches the corresponding number of current sources onto the current-voltage converter. The advantage of both techniques is the monotonicity which is guaranteed for ideal switches but also for slightly deviating resistances. The large number of resistors in switched current sources or the large number of switched current sources causes problems for long word-lengths. The techniques are used in combination with other methods for DA conversion of higher significant bits.

3.3.3   Weighted Resistors and Capacitors

A reduction in the number of identical resistors or current sources is achieved with the following method.

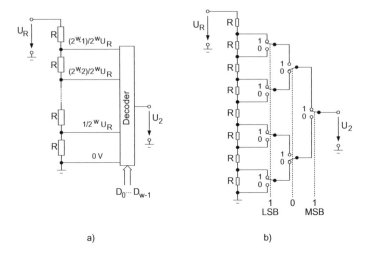

a)   b)

Figure 3.43 Switched voltage sources.

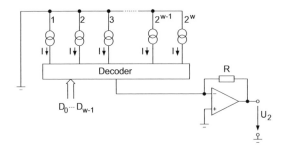

Figure 3.44 Switched current sources.

Weighted Resistors. DA conversion with w switched current sources which are weighted according to

$$I_1 = 2I_2 = 4I_3 = \cdots = 2^{w-1}I_w \tag{3.40}$$

is shown in Fig. 3.45. The output voltage is

$$U_2 = -R \cdot I = -R \cdot (b_1 I_1 2^0 + b_2 I_2 2^1 + b_3 I_3 2^2 + \cdots + b_w I_w 2^{w-1}), \tag{3.41}$$

where b_n takes values 0 or 1. The implementation of DA conversion with switched current sources is carried out with weighted resistors as shown in Fig. 3.46. The output voltage is

$$U_2 = R \cdot I = R\left(\frac{b_1}{2R} + \frac{b_2}{4R} + \frac{b_4}{8R} + \cdots + \frac{b_w}{2^w R}\right)U_R \tag{3.42}$$

$$= (b_1 2^{-1} + b_2 2^{-2} + b_3 2^{-3} + \cdots + b_w 2^{-w})U_R. \tag{3.43}$$

Weighted Capacitors. DA conversion with weighted capacitors is shown in Fig. 3.47. During the first phase (switch position 1 in Fig. 3.47) all capacitors are discharged. During

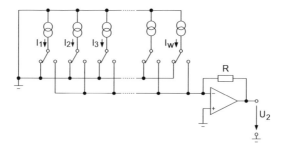

Figure 3.45 Weighted current sources.

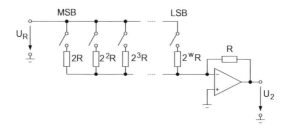

Figure 3.46 DA conversion with weighted resistors.

the second phase, all capacitors that belong to 1 bit are connected to a reference voltage. Those capacitors belonging to 0 bits are connected to ground. The charge on the capacitors C_a that are connected with the reference voltage can be set equal to the total charge on all capacitors C_g, which leads to

$$U_R\, C_a = U_R\left(b_1 C + \frac{b_2 C}{2} + \frac{b_3 C}{2^2} + \cdots + \frac{b_w C}{2^{w-1}}\right) = C_g\, U_2 = 2\, CU_2. \qquad (3.44)$$

Hence, the output voltage is

$$U_2 = (b_1 2^{-1} + b_2 2^{-2} + b_3 2^{-3} + \cdots + b_w 2^{-w})U_R. \qquad (3.45)$$

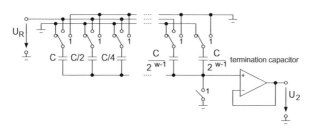

Figure 3.47 DA conversion with weighted capacitors.

3.3.4   R-2R Resistor Networks

The DA conversion with switched current sources can also be carried out with an R-2R resistor network as shown in Fig. 3.48. In contrast to the method with weighted resistors, the ratio of the smallest to largest resistor is reduced to 2:1.

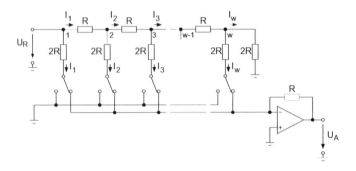

Figure 3.48 Switched current sources with R-2R resistor network.

The weighting of currents is achieved by a current division at every junction. Looking right from every junction, a resulting resistance $R + 2R \parallel 2R = 2R$ is found which is equal to the resistance in the vertical direction downwards from the junction. For the current from junction 1 it follows that $I_1 = U_R/2R$, and for the current from junction 2 $I_2 = I_1/2$. Hence, a binary weighting of the w currents is given by

$$I_1 = 2I_2 = 4I_3 = \cdots = 2^{w-1} I_w. \tag{3.46}$$

The output voltage U_2 can be written as

$$U_2 = -RI = -R\left(\frac{b_1}{2R} + \frac{b_2}{4R} + \frac{b_3}{8R} + \cdots + \frac{b_w}{2^{w-1}R}\right)U_R \tag{3.47}$$

$$= -U_R(b_1 2^{-1} + b_2 2^{-2} + b_3 2^{-3} + \cdots + b_w 2^{-w}). \tag{3.48}$$

3.3.5   Delta-sigma DA Converter

A delta-sigma DA converter is shown in Fig. 3.49. The converter is provided with w-bit data words by an input register with the sampling rate f_S. This is followed by a sample rate conversion up to Lf_S by upsampling and a digital low-pass filter. A delta-sigma modulator converts the w-bit input signal into a 1-bit output signal. The delta-sigma modulator corresponds to the model in Section 3.1.3. Subsequently, the DA conversion of the 1-bit signal is performed followed by the reconstruction of the time-continuous signal by an analog low-pass filter.

3.4   Java Applet – Oversampling and Quantization

The applet shown in Fig. 3.50 demonstrates the influence of oversampling on power spectral density of the quantization error. For a given quantization word-length the noise level

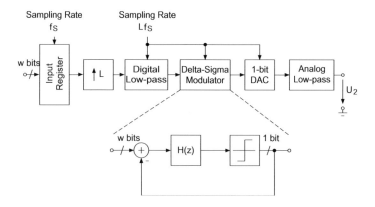

Figure 3.49 Delta-sigma DA converter.

can be reduced by changing the oversampling factor. The graphical interface of this applet presents several quantization and oversampling values; these can be used to experiment the noise reduction level. An additional FFT spectral representation provides a visualization of this audio effect.

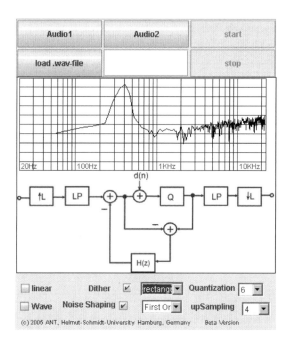

Figure 3.50 Java applet – oversampling and quantization.

The following functions can be selected on the lower right of the graphical user interface:

- Quantizer

  - word-length w leads to quantization step size $Q = 2^{w-1}$.

- Dither

  - rect dither – uniform probability density function
  - tri dither – triangular probability density function
  - high-pass dither – triangular probability density function and high-pass power spectral density.

- Noise shaping

  - first-order $H(z) = z^{-1}$.

- Oversampling factor

  - Factors from 4 up to 64 can be tested depending on the CPU performance of your machine.

You can choose between two predefined audio files from our web server (*audio1.wav* or *audio2.wav*) or your own local wav file to be processed [Gui05].

3.5   Exercises

1. Oversampling

1. How do we define the power spectral density $S_{XX}(e^{j\Omega})$ of a signal $x(n)$?

2. What is the relationship between signal power σ_X^2 (variance) and power spectral density $S_{XX}(e^{j\Omega})$?

3. Why do we need to oversample a time-domain signal?

4. Explain why an oversampled PCM A/D converter has lower quantization noise power in the base-band than a Nyquist rate sampled PCM A/D converter.

5. How do we perform oversampling by a factor of L in the time domain?

6. Explain the frequency-domain interpretation of the oversampling operation.

7. What is the pass-band and stop-band frequency of the analog anti-aliasing filter?

8. What is the pass-band and stop-band frequency of the digital anti-aliasing filter before downsampling?

9. How is the downsampling operation performed (time-domain and frequency-domain explanation)?

2. Delta-sigma Conversion

1. Why can we apply noise shaping in an oversampled AD converter?

2. Show how the delta-sigma converter (DSC) has a lower quantization error power in the base-band than an oversampled PCM A/D converter.

3. How do the power spectral density and variance change in relation to the order of the DSC?

4. How is noise shaping achieved in an oversampled delta-sigma AD converter?

5. Show the noise shaping effect (with Matlab plots) of a delta sigma modulator and how the improvement of the signal-to-noise for pure oversampling and delta-sigma modulator is achieved.

6. Using the previous Matlab plots, specify which order and oversampling factor L will be needed for a 1-bit delta-sigma converter for SNR $= 100$ dB.

7. What is the difference between the delta-sigma modulator in the delta-sigma AD converter and the delta-sigma DA converter?

8. How do we achieve a w-bit signal representation at Nyquist sampling frequency from an oversampled 1-bit signal?

9. Why do we need to oversample a w-bit signal for a delta-sigma DA converter?

References

[Can85]   J. C. Candy: *A Use of Double Integration in Sigma Delta Modulation*, IEEE Trans. Commun., Vol. COM-37, pp. 249–258, March 1985.

[Can92]   J. C. Candy, G. C. Temes, Ed.: *Oversampling Delta-Sigma Data Converters*, IEEE Press, Piscataway, NJ, 1992.

[Cha90]   K. Chao et al: *A High Order Topology for Interpolative Modulators for Oversampling A/D Converters*, IEEE Trans. Circuits and Syst., Vol. CAS-37, pp. 309–318, March 1990.

[Cro83]   R. E. Crochiere, L. R. Rabiner: *Multirate Digital Signal Processing*, Prentice Hall, Englewood Cliffs, NJ, 1983.

[Fli00]   N. Fliege: *Multirate Digital Signal Processing*, John Wiley & Sons, Ltd, Chichester, 2000.

[Gui05]   M. Guillemard, C. Ruwwe, U. Zölzer: *J-DAFx – Digital Audio Effects in Java*, Proc. 8th Int. Conference on Digital Audio Effects (DAFx-05), pp. 161–166, Madrid, 2005.

[Ino63]   H. Inose, Y. Yasuda: *A Unity Bit Coding Method by Negative Feedback*, Proc. IEEE, Vol. 51, pp. 1524–1535, November 1963.

[Mat87]   Y. Matsuya et al: *A 16-bit Oversampling A-to-D Conversion Technology Using Triple-Integration Noise Shaping*, IEEE J. Solid-State Circuits, Vol. SC-22, pp. 921–929, December 1987.

[She86]   D. H. Sheingold, Ed.: *Analog-Digital Conversion Handbook*, 3rd edn, Prentice Hall, Englewood Cliffs, NJ, 1986.

[Vai93]   P. P. Vaidyanathan: *Multirate Systems and Filter Banks*, Prentice Hall, Englewood Cliffs, NJ, 1993.

Chapter 4

Audio Processing Systems

Digital signal processors (DSPs) are used for discrete-time signal processing. Their architecture and instruction set is specially designed for real-time processing of signal processing algorithms. DSPs of different manufacturers and their use in practical circuits will be discussed. The restriction to the architecture and practical circuits will provide the user with the criteria necessary for selecting a DSP for a particular application. From the architectural features of different DSPs, the advantages of a certain processor with respect to fast execution of algorithms (digital filter, adaptive filter, FFT, etc.) automatically result. The programming methods and application programs are not dealt with here, because the DSP user guides from different manufacturers provide adequate information in the form of sample programs for a variety of signal processing algorithms.

After comparing DSPs with other microcomputers, the following topics will be discussed in the forthcoming sections:

- fixed-point DSPs;

- floating-point DSPs;

- development tools;

- single-processor systems (peripherals, control principles);

- multi-processor systems (coupling principles, control principles).

The internal design of microcomputers is mainly based on two architectures; the *von Neumann* architecture which uses a shared instruction/data bus; and the *Harvard* architecture which has separate buses for instructions and data. Processors based on these architectures are CISCs, RISCs and DSPs. Their characteristics are given in Table 4.1. Besides the internal properties listed in the table, DSPs have special on-chip peripherals which are suited to signal processing applications. The fast response to external interrupts enables their use in real-time operating systems.

Table 4.1 CISC, RISC and DSP.

Type	Characteristics
CISC	Complex Instruction Set Computer • von Neumann architecture • assembler programming • large number of instructions • computer families • compilers • application: universal microcomputers
RISC	Reduced Instruction Set Computer • von Neumann architecture/Harvard architecture • number of instructions <50 • number of address modes <4, instruction formats <4 • hard-wired instruction (no microprogramming) • processing most of the instructions in one cycle • optimizing compilers for high-level programming languages • application: workstations
DSP	Digital Signal Processor • Harvard architecture • several internal data buses • assembler programming • parallel processing of several instructions in one cycle • optimizing compilers for high-level programming languages • real-time operating systems • application: real-time signal processing

4.1   Digital Signal Processors

4.1.1   Fixed-point DSPs

The discrete-time and discrete-amplitude output of an AD converter is usually represented in 2's complement format. The processing of these number sequences is carried out with fixed-point or floating-point arithmetic. The output of a processed signal is again in 2's complement format and is fed to a DA converter. The signed fractional representation (2's complement) is the common method for algorithms in fixed-point number representation. For address generation and modulo operations unsigned integers are used. Figure 4.1 shows a schematic diagram of a typical fixed-point DSP. The main building blocks are program controller, arithmetic logic unit (ALU) with a multiplier-accumulator (MAC), program and data memory and interfaces to external memory and peripherals. All blocks are connected with each other by an internal bus system. The internal bus system has separate instruction and data buses. The data bus itself can consist of more than one parallel bus enabling it, for instance, to transmit both operands of a multiplication instruction to the MAC in parallel. The internal memory consists of instruction and data RAM and additional ROM memory. This internal memory permits fast execution of internal instructions and data transfer. For increasing memory space, address/control and data buses are connected to external memories like EPROM, ROM and RAM. The connection of the external bus system to the internal bus architecture has great influence on efficient execution of external instructions as well as on processing external data. In order to connect serially operating AD/DA converters, special serial interfaces with high transmission rates are offered by several DSPs. Moreover, some processors support direct connection to an RS232 interface. The control from a microprocessor can be achieved via a host interface with a word-length of 8 bits.

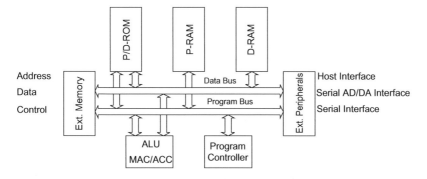

Figure 4.1 Schematic diagram of a fixed-point DSP.

An overview of fixed-point DSPs with respect to word-length and cycle time is shown in Table 4.2. Basically, the precision of the arithmetic can be doubled if quantization affects the stability and numeric precision of the applied algorithm. The cycle time in connection with processing time (in processor cycles) of a combined multiplication and accumulation command gives insight into the computing power of a particular processor type. The cycle time directly results from the maximum clock frequency. The instruction processing time depends mainly on the internal instruction and data structure as well as on the external memory connections of the processor. The fast access to external instruction and data memories is of special significance in complex algorithms and in processing huge data loads. Further attention has to be paid to the linking of serial data connections with AD/DA converters and the control by a host computer over a special host interface. Complex interface circuits could therefore be avoided. For stand-alone solutions, program loading from a simple external EPROM can also be done.

For signal processing algorithms, the following software commands are necessary:

1. MAC (multiply and accumulate) → combined multiplication and addition command;

2. simultaneous transfer of both operands for multiplication to the MAC (parallel move);

3. bit-reversed addressing (for FFT);

4. modulo addressing (for windowing and filtering).

Different signal processors have different processing times for FFT implementations. The latest signal processors with improved architecture have shorter processing times. The instruction cycles for the combined multiplication and accumulation command (application: windowing, filtering) are approximately equal for different processors, but processing cycles for external operands have to be considered.

Table 4.2 Fixed-point DSPs (Analog Devices AD, Texas Instruments TI, Motorola MOT, Agere Systems AG).

Type	Word-length	Cycle time MHz/ns	Computation power MMACS
ADSP-BF533	16	756/1.3	1512
ADSP-BF561	16	756/1.3	3024
ADSP-T201	32	600/1.67	4800
TI-TMS320C6414	16	1000/1	4000
MOT-DSP56309	24	100/10	100
MOT-DSP56L307	24	160/6.3	160
AG-DSP16410 × 2	16	195/5.1	780

4.1.2 Floating-point DSPs

Figure 4.2 shows the block diagram of a typical floating-point DSP. The main characteristics of the different architectures are the dual-port principle (Motorola, Texas Instruments) and the external Harvard architecture (Analog Devices, NEC). Floating-point DSPs internally have multiple bus systems in order to accelerate data transfer to the processing unit. An on-chip DMA controller and cache memory support higher data transfer rates. An overview of floating-point DSPs is shown in Table 4.3. Besides the standardized floating-point representation IEEE-754, there are also manufacturer-dependent number representations.

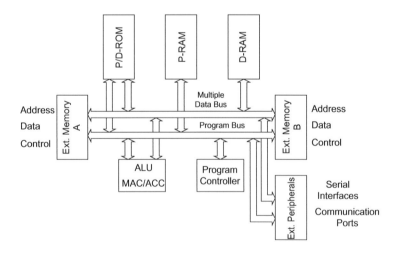

Figure 4.2 Block diagram of a floating-point digital signal processor.

Table 4.3 Floating-point DSPs.

Type	Word-length	Cycle time MHz/ns	Computation power MFLOPS
ADSP 21364	32	300/3.3	1800
ADSP 21267	32	150/6.6	900
ADSP-21161N	32	100/10	600
TI-TMS320C6711	32	200/5	1200

4.2 Digital Audio Interfaces

For transferring digital audio signals, two transmission standards have been established by the AES (Audio Engineering Society) and the EBU (European Broadcasting Union), respectively. These standards are for two-channel transmission [AES92] and for multichannel transmission of up to 56 audio signals.

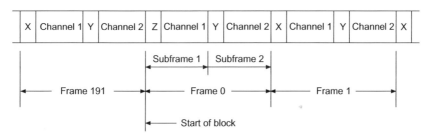

Figure 4.3 Two-channel format.

Figure 4.4 Two-channel format (subframe).

4.2.1 Two-channel AES/EBU Interface

For the two-channel AES/EBU interface, professional and consumer modes are defined. The outer frame is identical for both modes and is shown in Fig. 4.3. For a sampling period a frame is defined so that it consists of two subframes, for channel 1 with preamble X, and for channel 2 with preamble Y. A total of 192 frames form a block, and the block start is characterized by a special preamble Z. The bit allocation of a subframe consists of 32 bits as in Fig. 4.4. The preamble consists of 4 bits (bit 0, . . . , 3) and the audio data of up to 24 bits (bit 4, . . . , 27). The last four bits of the subframe characterize *Validity* (validity of data word or error), *User* Status (usable bit), *Channel* Status (from 192 bits/block = 24 bytes coded status information for the channel) and *Parity* (even parity). The transmission of the serial data bits is carried out with a biphase code. This is done with the help of an XOR relationship between clock (of double bit rate) and the serial data bits (Fig. 4.5). At the receiver, clock retrieval is achieved by detecting the preamble (X = 11100010, Y = 11100100, Z = 11101000) as it violates the coding rule (see Fig. 4.6). The meaning of the 24 bytes for channel status information is summarized in Table 4.4. An exact bit allocation of the first three important bytes of this channel status information is presented in Fig. 4.7. In the individual fields of byte 0, preemphasis and sampling rate are specified besides professional/consumer modes and the characterization of data/audio (see Tables 4.5 and 4.6). Byte 1 determines the channel mode (Table 4.7). The consumer format (often labeled SPDIF = Sony/Philips Digital Interface Format) differs from the professional format in the definition of the channel status information and the technical specifications for inputs and outputs. The bit allocation for the first four bits of the channel information is shown in Fig. 4.8. For consumer applications, two-wired leads with RCA connectors are used. The inputs and outputs are asymmetrical. Also, optical connectors exist. For professional use, shielded two-wired leads with XLR connectors and symmetrical inputs and outputs (professional format) are used. Table 4.8 shows the electrical specifications for professional AES/EBU interfaces.

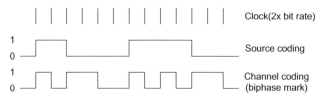

Figure 4.5 Channel coding.

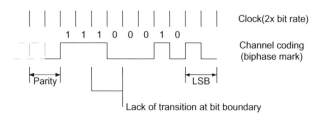

Figure 4.6 Preamble X.

4.2.2   MADI Interface

For connecting an audio processing system at different locations, a MADI interface (*M*ultichannel *A*udio *D*igital *I*nterface) is used. A system link by MADI is presented in Fig. 4.9. Analog/digital I/O systems consisting of AD/DA converters, AES/EBU interfaces (AES) and sampling rate converters (SRC) are connected to digital distribution systems with bi-directional MADI links. The actual audio signal processing is performed in special DSP systems which are connected to the digital distribution systems by MADI links. The

Table 4.4 Channel status bytes.

Byte	Description
0	Emphasis, sampling rate
1	Channel use
2	Sample length
3	Vector for byte 1
4	Reference bits
5	Reserved
6–9	4 bytes of ASCII origin
10–13	4 bytes of ASCII destination
14–17	4 bytes of local address
18–21	Time code
22	Flags
23	CRC

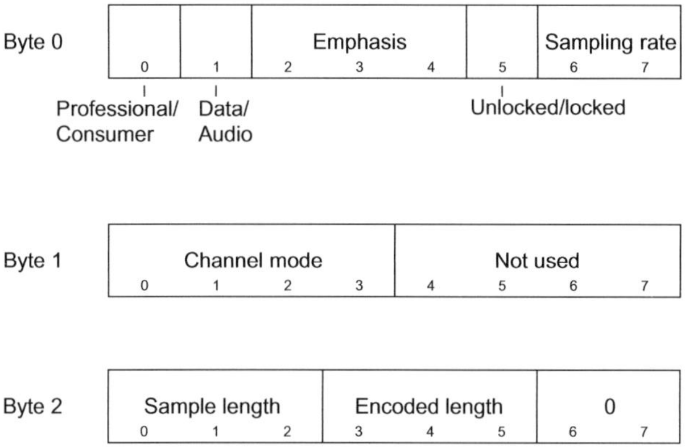

Figure 4.7 Bytes 0–2 of channel status information.

Table 4.5 Emphasis field.

0	None indicated, override enabled
4	None indicated, override disabled
6	50/15 μs emphasis
7	CCITT J.17 emphasis

Table 4.6 Sampling rate field.

0	None indicated (48 kHz default)
1	48 kHz
2	44.1 kHz
3	32 kHz

Table 4.7 Channel mode.

0	None indicated (2 channel default)
1	Two channel
2	Monaural
3	Primary/secondary (A = primary, B = secondary)
4	Stereo (A = left, B = right)
7	Vector to byte 3

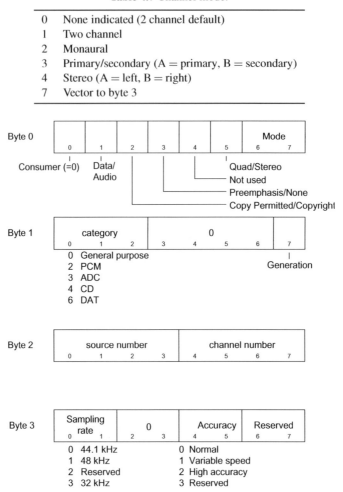

Figure 4.8 Bytes 0–3 (consumer format).

Table 4.8 Electrical specifications of professional interfaces.

Output impedance	Signal amplitude	Jitter
110 Ω	2–7 V	max. 20 ns
Input impedance	Signal amplitude	Connect.
110 Ω	min. 200 mV	XLR

MADI format is derived from the two-channel AES/EBU format and allows the transmission of 56 digital mono channels (see Fig. 4.10) within a sampling period. The MADI frame consists of 56 AES/EBU subframes. Each channel has a preamble containing the information shown in Fig. 4.10. The bit 0 is responsible for identifying the first MADI channel (MADI Channel 0). Table 4.9 shows the sampling rates and the corresponding data transfer rates. The maximum data rate of 96.768 Mbit/s is required at sampling rate of 48 kHz + 12.5%. Data transmission is done by FDDI techniques (*F*iber *D*istributed *D*igital *I*nterface). The transmission rate of 125 Mbit/s is implemented with special TAXI chips. The transmission for a coaxial cable is already specified (see Table 4.10). The optical transmission medium for audio applications is not yet defined.

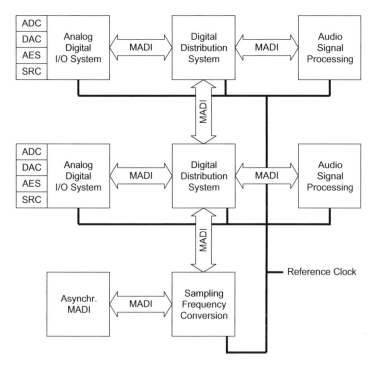

Figure 4.9 A system link by MADI.

Table 4.9 MADI specifications.

Sampling rate	32 kHz–48 kHz ± 12.5%
Transmission rate	125 Mbit/s
Data transfer rate	100 Mbit/s
Max. data transfer rate	96.768 Mbit/s (56 channels at 48 kHz + 12.5%)
Min. data transfer rate	50.176 Mbit/s (56 channels at 32 kHz − 12.5%)

A unidirectional MADI link is shown in Fig. 4.11. The MADI transmitter and receiver must be synchronized by a common master clock. The transmission between FDDI chips is performed by a transmitter with integrated clock generation and clock retrieval at the receiver.

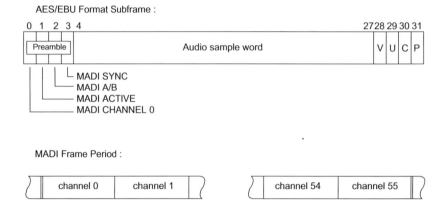

Figure 4.10 MADI frame format.

Table 4.10 Electrical specifications (MADI).

Output impedance	Signal ampl.	Cable length	Connect.
75 Ω	0.3–0.7 V	50 m (coaxial cable)	BNC

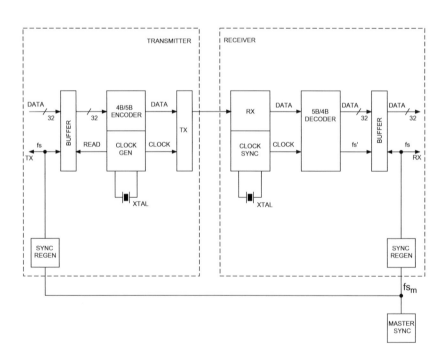

Figure 4.11 MADI link.

4.3 Single-processor Systems

4.3.1 Peripherals

A common system configuration is shown in Fig. 4.12. It consists of a DSP, clock generation, instruction and data memory and a BOOT-EPROM. After RESET, the program is
loaded into the internal RAM of the signal processor. The loading is done byte by byte
so that only an EPROM with 8-bit data word-length is necessary. In terms of circuit complexity the connection of AD/DA converters over serial interfaces is the simplest solution.
Most fixed-point signal processors support serial connection where a lead 'connection'
for bit clock SCLK, sampling clock/word clock WCLK, and the serial input and output
data SDRX/SDTX are used. The clock signals are obtained from a higher reference clock
CLKIN (see Fig. 4.13). For non-serially operating AD/DA converters, parallel interfaces
can also be connected to the DSP.

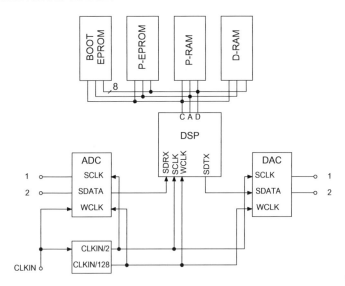

Figure 4.12 DSP system with two-channel AD/DA converters (C = control, A = address, D =
data, SDATA = serial data, SCLK = bit clock, WCLK = word clock, SDRX = serial input, SDTX =
serial output).

4.3.2 Control

For controlling digital signal processors and data exchange with host processors, some
DSPs provide a special host interface that can be read and written directly (see Fig. 4.14).
The data word-length depends on the processor. The host interface is included in the
external address space of the host or is connected to a local bus system, for instance a
PC bus.

  A DSP as a coprocessor for special signal processing problems can be used by connecting it with a dual-port RAM and additional interrupt logic to a host processor. This

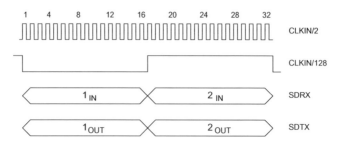

Figure 4.13 Serial transmission format.

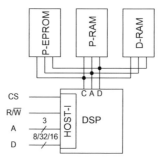

Figure 4.14 Control via a host interface of the DSP (CS = chip select, R/W = read/write, A = address, D = data).

enables data transmission between the DSP system and host processor (see Fig. 4.15). This results in a complete separation from the host processor. The communication can either be interrupt-controlled or carried out by polling a memory address in a dual-port RAM.

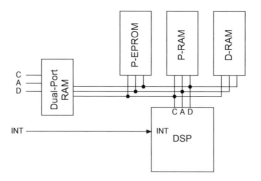

Figure 4.15 Control over a dual-port RAM and interrupt.

A very simple control can be done directly via an RS232-interface. This is can be carried out via an additional asynchronous serial interface (*Serial Communication Interface*) of the DSP (see Fig. 4.16).

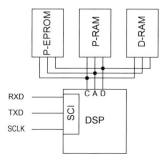

Figure 4.16 Control over a serial interface (RS232, RS422).

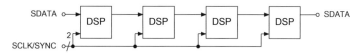

Figure 4.17 Cascading and pipelining (SDATA = serial data, SCLK = bit clock, SYNC = synchronization).

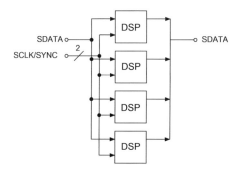

Figure 4.18 Parallel configuration with output time-multiplex.

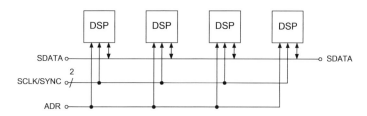

Figure 4.19 Time-multiplex connection (ADR = address at a particular time).

4.4 Multi-processor Systems

The design of multi-processor systems can be carried out by linking signal processors by serial or parallel interfaces. Besides purely multi-processor DSP systems, an additional connection to standard bus systems can be made as well.

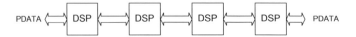

Figure 4.20 Cascading and pipelining.

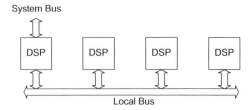

Figure 4.21 Parallel configuration.

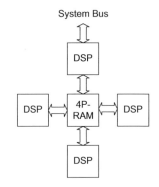

Figure 4.22 Connection over a four-port RAM.

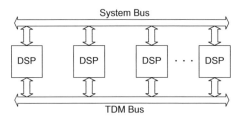

Figure 4.23 Signal processor systems based on standard bus system.

4.4.1   Connection via Serial Links

In connecting via serial links, signal processors are cascaded so that different program segments are distributed over different processors (see Fig. 4.17). The serial output data is fed into the serial input of the following signal processor. A synchronous bit clock and a common synchronization SYNC control the serial interface. With the help of a serial time-multiplex mode (Fig. 4.18) a parallel configuration can be designed which, for instance, feeds several parallel signal processors with serial input data. The serial outputs of signal

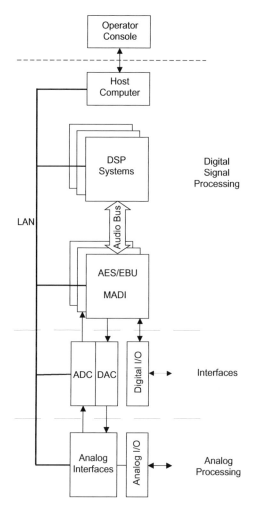

Figure 4.24 Audio system.

processors provide output data in time-multiplex. A complete time-multiplex connection via the serial interface of the signal processor is shown in Fig. 4.19. The allocation of a signal processor at a particular time slot can either be fixed or carried out by an address control ADR.

4.4.2 Connection via Parallel Links

Connection via parallel links is possible with dual-port processors as well as with dual-port RAMs (see Fig. 4.20). A parallel configuration of signal processor systems with a local bus is shown in Fig. 4.21. The connection to the local bus is done either over a dual-port RAM or directly with a second signal processor port. Another possible configuration is the use of a four-port RAM as shown in Fig. 4.22. Here, one processor serves as a connector to a

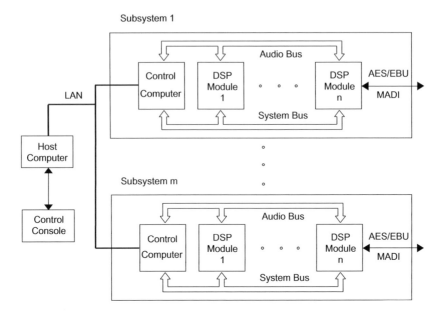

Figure 4.25 Scalable digital audio system.

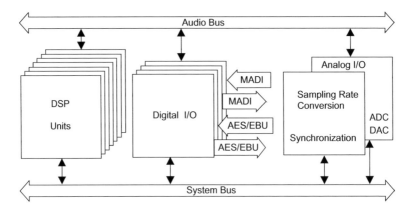

Figure 4.26 Subsystem.

system bus and feeds three other processors over a four-port RAM with control and data information.

4.4.3   Connection via Standard Bus Systems

The use of standard bus systems (VME bus, MULTIBUS, PC bus) to control multi-processor systems is presented in Fig. 4.23. The connection of signal processors can either be carried out directly over a control bus or with the help of a special data bus. This parallel data bus can operate in time-multiplex. Hence, control information and data are separated. A few of the criteria for standard bus systems are data transfer rate, interrupt request and processing, the option of several masters, auxiliary functions (power supply, bus error, battery buffer) and mechanical requirements.

4.4.4 Scalable Audio System

The functional segmentation of an audio system into different stages, the analog, interface, digital and man–machine stages, is shown in Fig. 4.24. All stages are controlled by a LAN (*Local Area Network*). In the analog domain, crosspoint switches and microphone amplifiers are controlled. In the interface domain AD/DA converters and sampling rate converters are used. The connection to a signal processing system is done by AES/EBU and MADI interfaces. A host computer with a control console for the sound engineer serves as the central control unit.

The realization of the digital domain with the help of a standard bus system is shown in Fig. 4.25. A central mixing console controls several subsystems over a host. These subsystems have special control computers which control several DSP modules. The system concept is scalable within a subsystem and by extension to several subsystems. Audio data transfer between subsystems is performed by AES/EBU and MADI interfaces. The segmentation within a subsystem is shown in Fig. 4.26. Here, besides DSP modules, digital interfaces (AES/EBU, MADI, sampling rate converters, etc.) and AD/DA converters can be integrated.

References

[AES91] AES10-1991 (ANSI S4.43-1991): AES Recommended Practice for Digital Audio Engineering – Serial Multichannel Audio Digital Interface (MADI).

[AES92] AES3-1992 (ANSI S4.40-1992): AES Recommended Practice for Digital Audio Engineering – Serial Transmission Format for Two-Channel Linearly Represented Digital Audio.

Chapter 5

Equalizers

Spectral sound equalization is one of the most important methods for processing audio signals. Equalizers are found in various forms in the transmission of audio signals from a sound studio to the listener. The more complex filter functions are used in sound studios. But in almost every consumer product like car radios, hifi amplifiers, simple filter functions are used for sound equalization. We first discuss basic filter types followed by the design and implementation of recursive audio filters. In Sections 5.3 and 5.4 linear phase nonrecursive filter structures and their implementation are introduced.

5.1 Basics

For filtering of audio signals the following filter types are used:

- **Low-pass and high-pass filters** with cutoff frequency f_c (3 dB cutoff frequency) are shown with their magnitude response in Fig. 5.1. They have a pass-band in the lower and higher frequency range, respectively.

- **Band-pass and band-stop filters** (magnitude responses in Fig. 5.1) have a center frequency f_c and a lower and upper f_l und f_u cutoff frequency. They have a pass- and stop-band in the middle of the frequency range. For the bandwidth of a band-pass or a band-stop filter we have

$$f_b = f_u - f_l. \tag{5.1}$$

Band-pass filters with a constant relative bandwidth f_b/f_c are very important for audio applications [Cre03]. The bandwidth is proportional to the center frequency, which is given by $f_c = \sqrt{f_l \cdot f_u}$ (see Fig. 5.2).

- **Octave filters** are band-pass filters with special cutoff frequencies given by

$$f_u = 2 \cdot f_l, \tag{5.2}$$
$$f_c = \sqrt{f_l \cdot f_u} = \sqrt{2} \cdot f_l. \tag{5.3}$$

A spectral decomposition of the audio frequency range with octave filters is shown in Fig. 5.3. At the lower and upper cutoff frequency an attenuation of -3 dB occurs.

Digital Audio Signal Processing   Second Edition   Udo Zölzer
© 2008 John Wiley & Sons, Ltd

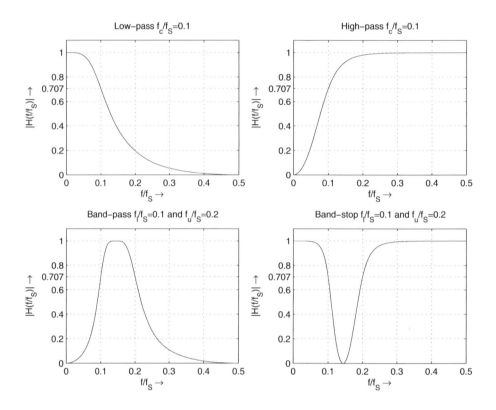

Figure 5.1 Linear magnitude responses of low-pass, high-pass, band-pass, and band-stop filters.

The upper octave band is represented as a high-pass. A parallel connection of octave filters can be used for a spectral analysis of the audio signal in octave frequency bands. This decomposition is used for the signal power distribution across the octave bands. For the center frequencies of octave bands we get $f_{c_i} = 2 \cdot f_{c_{i-1}}$. The weighting of octave bands with gain factors A_i and summation of the weighted octave bands represents an octave equalizer for sound processing (see Fig. 5.4). For this application the lower and upper cutoff frequencies need an attenuation of -6 dB, such that a sinusoid at the crossover frequency has gain of 0 dB. The attenuation of -6 dB is achieved through a series connection of two octave filters with -3 dB attenuation.

- **One-third octave filters** are band-pass filters (see Fig. 5.3) with cutoff frequencies given by

$$f_u = \sqrt[3]{2} \cdot f_l, \tag{5.4}$$
$$f_c = \sqrt[6]{2} \cdot f_l. \tag{5.5}$$

The attenuation at the lower and upper cutoff frequency is -3 dB. One-third octave filters split an octave into three frequency bands (see Fig. 5.3).

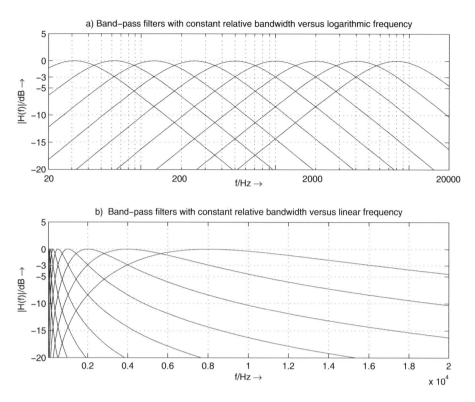

Figure 5.2 Logarithmic magnitude responses of band-pass filters with constant relative bandwidth.

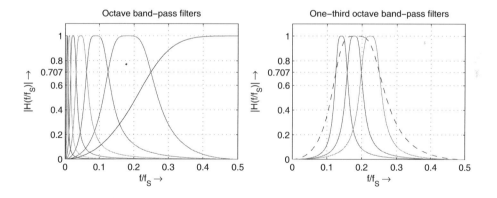

Figure 5.3 Linear magnitude responses of octave filters and decomposition of an octave band by three one-third octave filters.

- **Shelving filters** and **peak filters** are special weighting filters, which are based on low-pass/high-pass/band-pass filters and a direct path (see Section 5.2.2). They have no stop-band compared to low-pass/high-pass/band-pass filters. They are used in

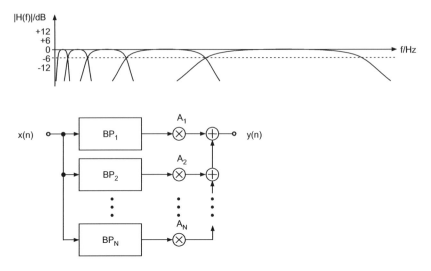

Figure 5.4 Parallel connection of band-pass filters (BP) for octave/one-third octave equalizers with gain factors (A_i for octave or one-third octave band).

a series connection of shelving and peak filters as shown in Fig. 5.5. The lower frequency range is equalized by low-pass shelving filters and the higher frequencies are modified by high-pass shelving filters. Both filter types allow the adjustment of cutoff frequency and gain factor. For the mid-frequency range a series connection of peak filters with variable center frequency, bandwidth, and gain factor are used. These shelving and peak filters can also be applied for octave and one-third octave equalizers in a series connection.

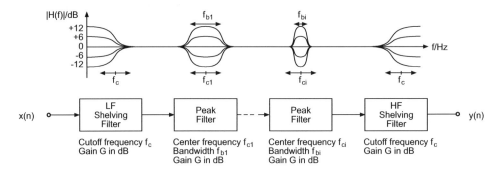

Figure 5.5 Series connection of shelving and peak filters (low-frequency LF, high-frequency HF).

- **Weighting filters** are used for signal level and noise measurement applications. The signal from a device under test is first passed through the weighting filter and then a root mean square or peak value measurement is performed. The two most often used filters are the A-weighting filter and the CCIR-468 weighting filter (see Fig. 5.6). Both weighting filters take the increased sensitivity of the human perception in the

1–6 kHz frequency range into account. The 0 dB of the magnitude response of both filters is crossed at 1 kHz. The CCIR-468 weighting filter has a gain of 12 dB at 6 kHz. A variant of the CCIR-468 filter is the ITU-ARM 2 kHz weighting filter, which is a 5.6 dB down tilted version of the CCIR-468 filters and passes the 0 dB at 2 kHz.

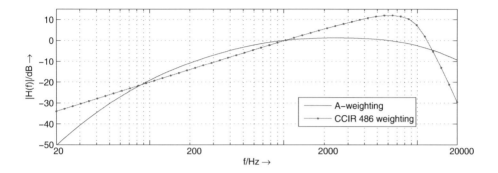

Figure 5.6 Magnitude responses of weighting filters for root mean square and peak value measurements.

5.2 Recursive Audio Filters

5.2.1 Design

A certain filter response can be approximated by two kinds of transfer function. On the one hand, the combination of poles and zeros leads to a very low-order transfer function $H(z)$ in fractional form, which solves the given approximation problem. The digital implementation of this transfer function needs recursive procedures owing to its poles. On the other hand, the approximation problem can be solved by placing only zeros in the z-plane. This transfer function $H(z)$ has, besides its zeros, a corresponding number of poles at the origin of the z-plane. The order of this transfer function, for the same approximation conditions, is substantially higher than for transfer functions consisting of poles and zeros. In view of an economical implementation of a filter algorithm in terms of computational complexity, recursive filters achieve shorter computing time owing to their lower order. For a sampling rate of 48 kHz, the algorithm has 20.83 μs processing time available. With the DSPs presently available it is easy to implement recursive digital filters for audio applications within this sampling period using only one DSP. To design the typical audio equalizers we will start with filter designs in the S-domain. These filters will then be mapped to the Z-domain by the bilinear transformation.

Low-pass/High-pass Filters. In order to limit the audio spectrum, low-pass and high-pass filters with Butterworth response are used in analog mixers. They offer a monotonic pass-band and a monotonically decreasing stop-band attenuation per octave ($n \cdot 6$ dB/oct.) that is determined by the filter order. Low-pass filters of the second and fourth order are commonly used. The normalized and denormalized second-order low-pass transfer functions are given by

$$H_{LP}(s) = \frac{1}{s^2 + \frac{1}{Q_\infty}s + 1} \quad \text{and} \quad H_{LP}(s) = \frac{\omega_c^2}{s^2 + \frac{\omega_c}{Q_\infty}s + \omega_c^2}, \tag{5.6}$$

where ω_c is the cutoff frequency and Q_∞ is the pole quality factor. The Q-factor Q_∞ of a Butterworth approximation is equal to $1/\sqrt{2}$. The denormalization of a transfer function is obtained by replacing the Laplace variable s by s/ω_g in the normalized transfer function.

The corresponding second-order high-pass transfer functions

$$H_{HP}(s) = \frac{s^2}{s^2 + \frac{1}{Q_\infty}s + 1} \quad \text{and} \quad H_{HP}(s) = \frac{s^2}{s^2 + \frac{\omega_c}{Q_\infty}s + \omega_c^2} \tag{5.7}$$

are obtained by a low-pass to high-pass transformation. Figure 5.7 shows the pole-zero locations in the s-plane. The amplitude frequency response of a high-pass filter with a 3 dB cutoff frequency of 50 Hz and a low-pass filter with a 3 dB cutoff frequency of 5000 Hz are shown in Fig. 5.8. Second- and fourth-order filters are shown.

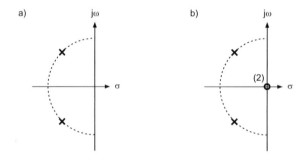

Figure 5.7 Pole-zero location for (a) second-order low-pass and (b) second-order high-pass.

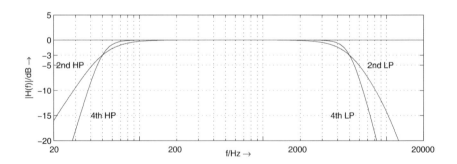

Figure 5.8 Frequency response of low-pass and high-pass filters – high-pass $f_c = 50\,\text{Hz}$ (second/fourth order), low-pass $f_c = 5000\,\text{Hz}$ (second/fourth order).

Table 5.1 summarizes the transfer functions of low-pass and high-pass filters with Butterworth response.

Table 5.1 Transfer functions of low-pass and high-pass filters.

Low-pass	$H(s) = \dfrac{1}{s^2 + \sqrt{2}s + 1}$	second order
	$H(s) = \dfrac{1}{(s^2 + 1.848s + 1)(s^2 + 0.765s + 1)}$	fourth order
High-pass	$H(s) = \dfrac{s^2}{s^2 + \sqrt{2}s + 1}$	second order
	$H(s) = \dfrac{s^4}{(s^2 + 1.848s + 1)(s^2 + 0.765s + 1)}$	fourth order

Band-pass and band-stop filters. The normalized and denormalized band-pass transfer functions of second order are

$$H_{BP}(s) = \frac{\frac{1}{Q_\infty}s}{s^2 + \frac{1}{Q_\infty}s + 1} \quad \text{and} \quad H_{BP}(s) = \frac{\frac{\omega_c}{Q_\infty}s}{s^2 + \frac{\omega_c}{Q_\infty}s + \omega_c^2}, \tag{5.8}$$

and the band-stop transfer functions are given by

$$H_{BS}(s) = \frac{s^2 - 1}{s^2 + \frac{1}{Q_\infty}s + 1} \quad \text{and} \quad H_{BS}(s) = \frac{s^2 - \omega_c^2}{s^2 + \frac{\omega_c}{Q_\infty}s + \omega_c^2}. \tag{5.9}$$

The relative bandwidth can be expressed by the Q-factor

$$Q_\infty = \frac{f_c}{f_b}, \tag{5.10}$$

which is the ratio of center frequency f_c and the 3 dB bandwidth given by f_b. The magnitude responses of band-pass filters with constant relative bandwidth are shown in Fig. 5.2. Such filters are also called *constant-Q filters*. The geometric symmetric behavior of the frequency response regarding the center frequency f_c is clearly noticeable (symmetry regarding the center frequency using a logarithmic frequency axis).

Shelving Filters. Besides the purely band-limiting filters like low-pass and high-pass filters, shelving filters are used to perform weighting of certain frequencies. A simple approach for a first-order low-pass shelving filter is given by

$$H(s) = 1 + H_{LP}(s) = 1 + \frac{H_0}{s + 1}. \tag{5.11}$$

It consists of a first-order low-pass filter with dc amplification of H_0 connected in parallel with an all-pass system of transfer function equal to 1. Equation (5.11) can be written as

$$H(s) = \frac{s + (1 + H_0)}{s + 1} = \frac{s + V_0}{s + 1}, \tag{5.12}$$

where V_0 determines the amplification at $\omega = 0$. By changing the parameter V_0, any desired boost ($V_0 > 1$) and cut ($V_0 < 1$) level can be adjusted. Figure 5.9 shows the frequency responses for $f_c = 100$ Hz. For $V_0 < 1$, the cutoff frequency is dependent on V_0 and is moved toward lower frequencies.

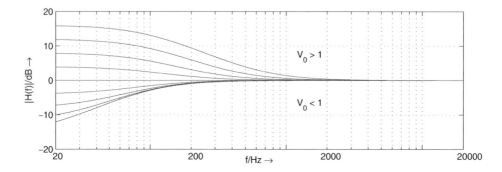

Figure 5.9 Frequency response of transfer function (5.12) with varying V_0 and cutoff frequency $f_c = 100$ Hz.

In order to obtain a symmetrical frequency response with respect to the zero decibel line without changing the cutoff frequency, it is necessary to invert the transfer function (5.12) in the case of cut ($V_0 < 1$). This has the effect of swapping poles with zeros and leads to the transfer function

$$H(s) = \frac{s+1}{s+V_0} \tag{5.13}$$

for the cut case. Figure 5.10 shows the corresponding frequency responses for varying V_0.

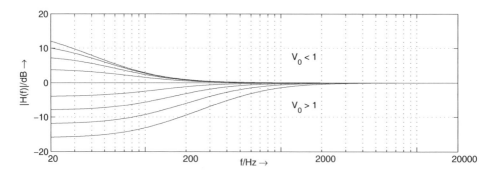

Figure 5.10 Frequency responses of transfer function (5.13) with varying V_0 and cutoff frequency $f_c = 100$ Hz.

Finally, Figure 5.11 shows the locations of poles and zeros for both the boost and the cut case. By moving zeros and poles on the negative σ-axis, boost and cut can be adjusted.

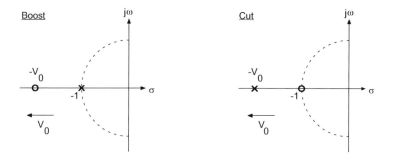

Figure 5.11 Pole-zero locations of a first-order low-frequency shelving filter.

The equivalent shelving filter for high frequencies can be obtained by

$$H(s) = 1 + H_{HP}(s) = 1 + \frac{H_0 s}{s + 1}, \qquad (5.14)$$

which is a parallel connection of a first-order high-pass with gain H_0 and a system with transfer function equal to 1. In the boost case the transfer function can written with $V_0 = H_0 + 1$ as

$$H(s) = \frac{s V_0 + 1}{s + 1}, \qquad V_0 > 1, \qquad (5.15)$$

and for cut we get

$$H(s) = \frac{s + 1}{s V_0 + 1}, \qquad V_0 > 1. \qquad (5.16)$$

The parameter V_0 determines the value of the transfer function $H(s)$ at $\omega = \infty$ for high-frequency shelving filters.

In order to increase the slope of the filter response in the transition band, a general second-order transfer function

$$H(s) = \frac{a_2 s^2 + a_1 s + a_0}{s^2 + \sqrt{2}s + 1} \qquad (5.17)$$

is considered, in which complex zeros are added to the complex poles. The calculation of poles leads to

$$s_{\infty\,1/2} = \sqrt{\tfrac{1}{2}}(-1 \pm j). \qquad (5.18)$$

If the complex zeros

$$s_{0\,1/2} = \sqrt{\frac{V_0}{2}}(-1 \pm j) \qquad (5.19)$$

are moved on a straight line with the help of the parameter V_0 (see Fig. 5.12), the transfer function

$$H(s) = \frac{s^2 + \sqrt{2 V_0}s + V_0}{s^2 + \sqrt{2}s + 1} \qquad (5.20)$$

of a second-order low-frequency shelving filter is obtained. The parameter V_0 determines the boost for low frequencies. The cut case can be achieved by inversion of (5.20).

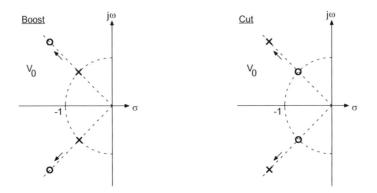

Figure 5.12 Pole-zero locations of a second-order low-frequency shelving filter.

A low-pass to high-pass transformation of (5.20) provides the transfer function

$$H(s) = \frac{V_0 s^2 + \sqrt{2V_0}s + 1}{s^2 + \sqrt{2}s + 1} \qquad (5.21)$$

of a second-order high-frequency shelving filter. The zeros

$$s_{\circ\,1/2} = \sqrt{\frac{1}{2V_0}}(-1 \pm j) \qquad (5.22)$$

are moved on a straight line toward the origin with increasing V_0 (see Fig. 5.13). The cut case is obtained by inverting the transfer function (5.21). Figure 5.14 shows the amplitude frequency response of a second-order low-frequency shelving filter with cutoff frequency 100 Hz and a second-order high-frequency shelving filter with cutoff frequency 5000 Hz (parameter V_0).

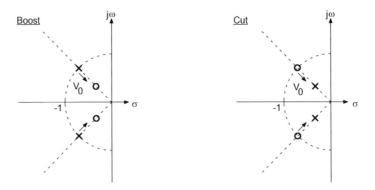

Figure 5.13 Pole-zero locations of second-order high-frequency shelving filter.

Peak Filter. Another equalizer used for boosting or cutting any desired frequency is the peak filter. A peak filter can be obtained by a parallel connection of a direct path and a

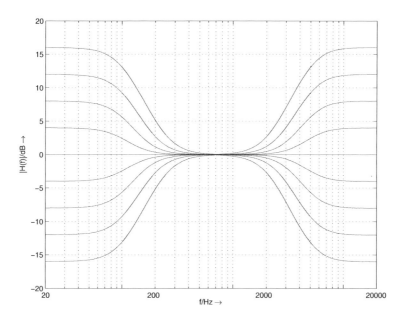

Figure 5.14 Frequency responses of second-order low-/high-frequency shelving filters – low-frequency shelving filter $f_c = 100$ Hz (parameter V_0), high-frequency shelving filter $f_c = 5000$ Hz (parameter V_0).

band-pass according to

$$\boxed{H(s) = 1 + H_{BP}(s).}$$ (5.23)

With the help of a second-order band-pass transfer function

$$H_{BP}(s) = \frac{(H_0/Q_\infty)s}{s^2 + \frac{1}{Q_\infty}s + 1},$$ (5.24)

the transfer function

$$H(s) = 1 + H_{BP}(s) = \frac{s^2 + \frac{1+H_0}{Q_\infty}s + 1}{s^2 + \frac{1}{Q_\infty}s + 1} = \frac{s^2 + \frac{V_0}{Q_\infty}s + 1}{s^2 + \frac{1}{Q_\infty}s + 1}$$ (5.25)

of a peak filter can be derived. It can be shown that the maximum of the amplitude frequency response at the center frequency is determined by the parameter V_0. The relative bandwidth is fixed by the Q-factor. The geometrical symmetry of the frequency response relative to the center frequency remains constant for the transfer function of a peak filter (5.25). The poles and zeros lie on the unit circle. By adjusting the parameter V_0, the complex zeros are moved with respect to the complex poles. Figure 5.15 shows this for the boost and cut cases. With increasing Q-factor, the complex poles move toward the $j\omega$-axis on the unit circle.

Figure 5.16 shows the amplitude frequency response of a peak filter by changing the parameter V_0 at a center frequency of 500 Hz and a Q-factor of 1.25. Figure 5.17 shows

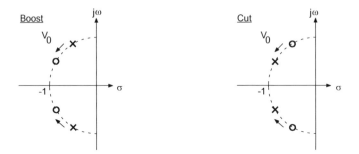

Figure 5.15 Pole-zero locations of a second-order peak filter.

the variation of the Q-factor Q_∞ at a center frequency of 500 Hz, a boost/cut of ± 16 dB and Q-factor of 1.25. Finally, the variation of the center frequency with boost and cut of ± 16 dB and a Q-factor 1.25 is shown in Fig. 5.18.

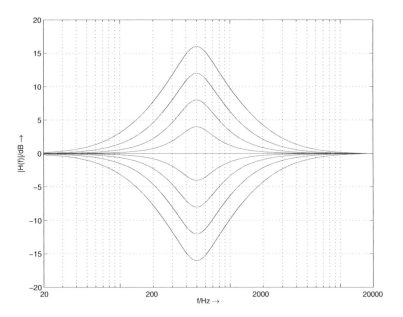

Figure 5.16 Frequency response of a peak filter – $f_c = 500$ Hz, $Q_\infty = 1.25$, cut parameter V_0.

Mapping to Z-domain. In order to implement a digital filter, the filter designed in the S-domain with transfer function $H(s)$ is converted to the Z-domain with the help of a suitable transformation to obtain the transfer function $H(z)$. The impulse-invariant transformation is not suitable as it leads to overlapping effects if the transfer function $H(s)$ is not band-limited to half the sampling rate. An independent mapping of poles and zeros from the S-domain into poles and zeros in the Z-domain is possible with help of the bilinear

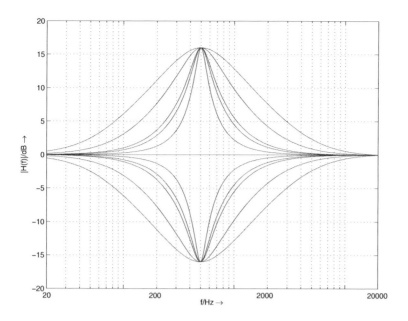

Figure 5.17 Frequency responses of peak filters – $f_c = 500$ Hz, boost/cut ± 16 dB, $Q_\infty = 0.707$, 1.25, 2.5, 3, 5.

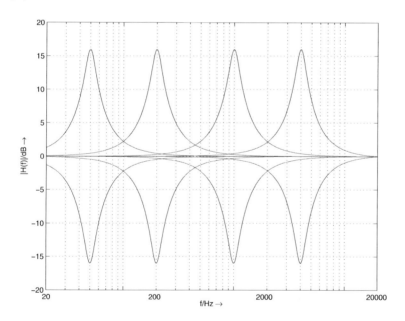

Figure 5.18 Frequency responses of peak filters – boost/cut ± 16 dB, $Q_\infty = 1.25$, $f_c = 50$, 200, 1000, 4000 Hz.

transformation given by

$$s = \frac{2}{T}\frac{z-1}{z+1}.$$ (5.26)

Tables 5.2–5.5 contain the coefficients of the second-order transfer function

$$H(z) = \frac{a_0 + a_1 z^{-1} + a_2 z^{-2}}{1 + b_1 z^{-1} + b_2 z^{-2}},$$ (5.27)

which are determined by the bilinear transformation and the auxiliary variable $K = \tan(\omega_c T/2)$ for all audio filter types discussed. Further filter designs of peak and shelving filters are discussed in [Moo83, Whi86, Sha92, Bri94, Orf96a, Dat97, Cla00]. A method for reducing the warping effect of the bilinear transform is proposed in [Orf96b]. Strategies for time-variant switching of audio filters can be found in [Rab88, Mou90, Zöl93, Din95, Väl98].

Table 5.2 Low-pass/high-pass/band-pass filter design.

Low-pass (second order)				
a_0	a_1	a_2	b_1	b_2
$\dfrac{K^2}{1+\sqrt{2}K+K^2}$	$\dfrac{2K^2}{1+\sqrt{2}K+K^2}$	$\dfrac{K^2}{1+\sqrt{2}K+K^2}$	$\dfrac{2(K^2-1)}{1+\sqrt{2}K+K^2}$	$\dfrac{1-\sqrt{2}K+K^2}{1+\sqrt{2}K+K^2}$
High-pass (second order)				
a_0	a_1	a_2	b_1	b_2
$\dfrac{1}{1+\sqrt{2}K+K^2}$	$\dfrac{-2}{1+\sqrt{2}K+K^2}$	$\dfrac{1}{1+\sqrt{2}K+K^2}$	$\dfrac{2(K^2-1)}{1+\sqrt{2}K+K^2}$	$\dfrac{1-\sqrt{2}K+K^2}{1+\sqrt{2}K+K^2}$
Band-pass (second order)				
a_0	a_1	a_2	b_1	b_2
$\dfrac{\frac{1}{Q}K}{1+\frac{1}{Q}K+K^2}$	0	$-\dfrac{\frac{1}{Q}K}{1+\frac{1}{Q}K+K^2}$	$\dfrac{2(K^2-1)}{1+\frac{1}{Q}K+K^2}$	$\dfrac{1-\frac{1}{Q}K+K^2}{1+\frac{1}{Q}K+K^2}$

5.2.2 Parametric Filter Structures

Parametric filter structures allow direct access to the parameters of the transfer function, like center/cutoff frequency, bandwidth and gain, via control of associated coefficients. To modify one of these parameters, it is therefore not necessary to compute a complete set of coefficients for a second-order transfer function, but instead only one coefficient in the filter structure is calculated.

Independent control of gain, cutoff/center frequency and bandwidth for shelving and peak filters is achieved by a feed forward (FF) structure for *boost* and a feed backward

Table 5.3 Peak filter design.

a_0	a_1	a_2	b_1	b_2
Peak (boost $V_0 = 10^{G/20}$)				
$\dfrac{1 + \frac{V_0}{Q_\infty}K + K^2}{1 + \frac{1}{Q_\infty}K + K^2}$	$\dfrac{2(K^2 - 1)}{1 + \frac{1}{Q_\infty}K + K^2}$	$\dfrac{1 - \frac{V_0}{Q_\infty}K + K^2}{1 + \frac{1}{Q_\infty}K + K^2}$	$\dfrac{2(K^2 - 1)}{1 + \frac{1}{Q_\infty}K + K^2}$	$\dfrac{1 - \frac{1}{Q_\infty}K + K^2}{1 + \frac{1}{Q_\infty}K + K^2}$
Peak (cut $V_0 = 10^{-G/20}$)				
$\dfrac{1 + \frac{1}{Q_\infty}K + K^2}{1 + \frac{V_0}{Q_\infty}K + K^2}$	$\dfrac{2(K^2 - 1)}{1 + \frac{V_0}{Q_\infty}K + K^2}$	$\dfrac{1 - \frac{1}{Q_\infty}K + K^2}{1 + \frac{V_0}{Q_\infty}K + K^2}$	$\dfrac{2(K^2 - 1)}{1 + \frac{V_0}{Q_\infty}K + K^2}$	$\dfrac{1 - \frac{V_0}{Q_\infty}K + K^2}{1 + \frac{V_0}{Q_\infty}K + K^2}$

Table 5.4 Low-frequency shelving filter design.

a_0	a_1	a_2	b_1	b_2
Low-frequency shelving (boost $V_0 = 10^{G/20}$)				
$\dfrac{1 + \sqrt{2V_0}K + V_0K^2}{1 + \sqrt{2}K + K^2}$	$\dfrac{2(V_0K^2 - 1)}{1 + \sqrt{2}K + K^2}$	$\dfrac{1 - \sqrt{2V_0}K + V_0K^2}{1 + \sqrt{2}K + K^2}$	$\dfrac{2(K^2 - 1)}{1 + \sqrt{2}K + K^2}$	$\dfrac{1 - \sqrt{2}K + K^2}{1 + \sqrt{2}K + K^2}$
Low-frequency shelving (cut $V_0 = 10^{-G/20}$)				
$\dfrac{1 + \sqrt{2}K + K^2}{1 + \sqrt{2V_0}K + V_0K^2}$	$\dfrac{2(K^2 - 1)}{1 + \sqrt{2V_0}K + V_0K^2}$	$\dfrac{1 - \sqrt{2}K + K^2}{1 + \sqrt{2V_0}K + V_0K^2}$	$\dfrac{2(V_0K^2 - 1)}{1 + \sqrt{2V_0}K + V_0K^2}$	$\dfrac{1 - \sqrt{2V_0}K + V_0K^2}{1 + \sqrt{2V_0}K + V_0K^2}$

Table 5.5 High-frequency shelving filter design.

a_0	a_1	a_2	b_1	b_2
High-frequency shelving (boost $V_0 = 10^{G/20}$)				
$\dfrac{V_0 + \sqrt{2V_0}K + K^2}{1 + \sqrt{2}K + K^2}$	$\dfrac{2(K^2 - V_0)}{1 + \sqrt{2}K + K^2}$	$\dfrac{V_0 - \sqrt{2V_0}K + K^2}{1 + \sqrt{2}K + K^2}$	$\dfrac{2(K^2 - 1)}{1 + \sqrt{2}K + K^2}$	$\dfrac{1 - \sqrt{2}K + K^2}{1 + \sqrt{2}K + K^2}$
High-frequency shelving (cut $V_0 = 10^{-G/20}$)				
$\dfrac{1 + \sqrt{2}K + K^2}{V_0 + \sqrt{2V_0}K + K^2}$	$\dfrac{2(K^2 - 1)}{V_0 + \sqrt{2V_0}K + K^2}$	$\dfrac{1 - \sqrt{2}K + K^2}{V_0 + \sqrt{2V_0}K + K^2}$	$\dfrac{2(K^2/V_0 - 1)}{1 + \sqrt{2/V_0}K + K^2/V_0}$	$\dfrac{1 - \sqrt{2/V_0}K + K^2/V_0}{1 + \sqrt{2/V_0}K + K^2/V_0}$

(FB) structure for *cut* as shown in Fig. 5.19. The corresponding transfer functions are:

$$G_{FW}(z) = 1 + H_0 H(z), \tag{5.28}$$

$$G_{FB}(z) = \frac{1}{1 + H_0 H(z)}. \tag{5.29}$$

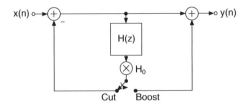

Figure 5.19 Filter structure for implementing boost and cut filters.

The boost/cut factor is $V_0 = 1 + H_0$. For digital filter implementations, it is necessary for the FB case that the inner transfer function be of the form $H(z) = z^{-1}H_1(z)$ to ensure causality. A parametric filter structure proposed by Harris [Har93] is based on the FF/FB technique, but the frequency response shows slight deviations near $z = 1$ and $z = -1$ from that desired. This is due to the z^{-1} in the FF/FB branch. Delay-free loops inside filter computations can be solved by the methods presented in [Här98, Fon01, Fon03]. Higher-order parametric filter designs have been introduced in [Kei04, Orf05, Hol06a, Hol06b, Hol06c, Hol06d]. It is possible to implement typical audio filters with only an FF structure. The complete decoupling of the control parameters is possible for the boost case, but there remains a coupling between bandwidth and gain factor for the cut case. In the following, two approaches for parametric audio filter structures based on an all-pass decomposition of the transfer function will be discussed.

Regalia Filter [Reg87]. The denormalized transfer function of a first-order shelving filter is given by

$$H(s) = \frac{s + V_0\omega_c}{s + \omega_c} \tag{5.30}$$

with

$$H(0) = V_0,$$
$$H(\infty) = 1.$$

A decomposition of (5.30) leads to

$$H(s) = \frac{s}{s + \omega_c} + V_0\frac{\omega_c}{s + \omega_c}. \tag{5.31}$$

The low-pass and high-pass transfer functions in (5.31) can be expressed by an all-pass decomposition of the form

$$\frac{s}{s + \omega_c} = \frac{1}{2}\left[1 + \frac{s - \omega_c}{s + \omega_c}\right], \tag{5.32}$$

$$\frac{V_0\omega_c}{s + \omega_c} = \frac{V_0}{2}\left[1 - \frac{s - \omega_c}{s + \omega_c}\right]. \tag{5.33}$$

With the all-pass transfer function

$$A_B(s) = \frac{s - \omega_c}{s + \omega_c} \tag{5.34}$$

for *boost*, (5.30) can be rewritten as

$$H(s) = \tfrac{1}{2}[1 + A_B(s)] + \tfrac{1}{2}V_0[1 - A_B(s)].$$ (5.35)

The bilinear transformation

$$s = \frac{2}{T}\frac{z-1}{z+1}$$

leads to

$$H(z) = \tfrac{1}{2}[1 + A_B(z)] + \tfrac{1}{2}V_0[1 - A_B(z)]$$ (5.36)

with

$$A_B(z) = -\frac{z^{-1} + a_B}{1 + a_B z^{-1}}$$ (5.37)

and the frequency parameter

$$a_B = \frac{\tan(\omega_c T/2) - 1}{\tan(\omega_c T/2) + 1}.$$ (5.38)

A filter structure for direct implementation of (5.36) is presented in Fig. 5.20a. Other possible structures can be seen in Fig. 5.20b,c. For the cut case $V_0 < 1$, the cutoff frequency of the filter moves toward lower frequencies [Reg87].

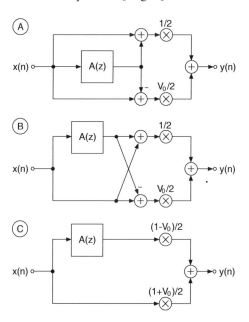

Figure 5.20 Filter structures by Regalia.

In order to retain the cutoff frequency for the cut case [Zöl95], the denormalized transfer function of a first-order shelving filter (cut)

$$H(s) = \frac{s + \omega_c}{s + \omega_c/V_0},$$ (5.39)

with the boundary conditions

$$H(0) = V_0,$$
$$H(\infty) = 1,$$

can be decomposed as

$$H(s) = \frac{s}{s + \omega_c/V_0} + \frac{\omega_c}{s + \omega_c/V_0}. \tag{5.40}$$

With the all-pass decompositions

$$\frac{s}{s + \omega_c/V_0} = \frac{1}{2}\left[1 + \frac{s - \omega_c/V_0}{s + \omega_c/V_0}\right], \tag{5.41}$$

$$\frac{\omega_c}{s + \omega_c/V_0} = \frac{V_0}{2}\left[1 - \frac{s - \omega_c/V_0}{s + \omega_c/V_0}\right], \tag{5.42}$$

and the all-pass transfer function

$$A_C(s) = \frac{s - \omega_c/V_0}{s + \omega_c/V_0} \tag{5.43}$$

for *cut*, (5.39) can be rewritten as

$$H(s) = \frac{1}{2}[1 + A_C(s)] + \frac{V_0}{2}[1 - A_C(s)]. \tag{5.44}$$

The bilinear transformation leads to

$$H(z) = \frac{1}{2}[1 + A_C(z)] + \frac{V_0}{2}[1 - A_C(z)] \tag{5.45}$$

with

$$A_C(z) = -\frac{z^{-1} + a_C}{1 + a_C z^{-1}} \tag{5.46}$$

and the frequency parameter

$$a_C = \frac{\tan(\omega_c T/2) - V_0}{\tan(\omega_c T/2) + V_0}. \tag{5.47}$$

Due to (5.45) and (5.36), boost and cut can be implemented with the same filter structure (see Fig. 5.20). However, it has to be noted that the frequency parameter a_C as in (5.47) for cut depends on the cutoff frequency and gain.

A second-order peak filter is obtained by a low-pass to band-pass transformation according to

$$z^{-1} \rightarrow -z^{-1}\frac{z^{-1} + d}{1 + dz^{-1}}. \tag{5.48}$$

For an all-pass as given in (5.37) and (5.46), the second-order all-pass is given by

$$A_{BC}(z) = \frac{z^{-2} + d(1 + a_{BC})z^{-1} + a_{BC}}{1 + d(1 + a_{BC})z^{-1} + a_{BC}z^{-2}} \tag{5.49}$$

with parameters (cut as in [Zöl95])

$$d = -\cos(\Omega_c), \tag{5.50}$$

$$V_0 = H(e^{j\Omega_c}), \tag{5.51}$$

$$a_B = \frac{1 - \tan(\omega_b T/2)}{1 + \tan(\omega_b T/2)}, \tag{5.52}$$

$$a_C = \frac{V_0 - \tan(\omega_b T/2)}{V_0 + \tan(\omega_b T/2)}. \tag{5.53}$$

The center frequency f_c is fixed by the parameter d, the bandwidth f_b by the parameters a_B and a_C, and gain by the parameter V_0.

Simplified All-pass Decomposition [Zöl95]. The transfer function of a first-order low-frequency shelving filter can be decomposed as

$$H(s) = \frac{s + V_0\omega_c}{s + \omega_c}$$

$$= 1 + H_0 \frac{\omega_c}{s + \omega_c} \tag{5.54}$$

$$= 1 + \frac{H_0}{2}\left[1 - \frac{s - \omega_c}{s + \omega_c}\right] \tag{5.55}$$

with

$$V_0 = H(s = 0), \tag{5.56}$$

$$H_0 = V_0 - 1, \tag{5.57}$$

$$V_0 = 10^{G/20} \quad \text{(G in dB)}. \tag{5.58}$$

The transfer function (5.55) is composed of a direct branch and a low-pass filter. The first-order low-pass filter is again implemented by an all-pass decomposition. Applying the bilinear transformation to (5.55) leads to

$$H(z) = 1 + \frac{H_0}{2}[1 - A(z)] \tag{5.59}$$

with

$$A(z) = -\frac{z^{-1} + a_B}{1 + a_B z^{-1}}. \tag{5.60}$$

For cut, the following decomposition can be derived:

$$H(s) = \frac{s + \omega_c}{s + \omega_c/V_0} \tag{5.61}$$

$$= 1 + \underbrace{(V_0 - 1)}_{H_0} \frac{\omega_c/V_0}{s + \omega_c/V_0} \tag{5.62}$$

$$= 1 + \frac{H_0}{2}\left[1 - \frac{s - \omega_c/V_0}{s + \omega_c/V_0}\right]. \tag{5.63}$$

The bilinear transformation applied to (5.63) again gives (5.59). The filter structure is identical for boost and cut. The frequency parameter a_B for boost and a_C for cut can be calculated as

$$a_B = \frac{\tan(\omega_c T/2) - 1}{\tan(\omega_c T/2) + 1},$$
(5.64)

$$a_C = \frac{\tan(\omega_c T/2) - V_0}{\tan(\omega_c T/2) + V_0}.$$
(5.65)

The transfer function of a first-order low-frequency shelving filter can be calculated as

$$H(z) = \frac{1 + (1 + a_{BC})\frac{H_0}{2} + (a_{BC} + (1 + a_{BC})\frac{H_0}{2})z^{-1}}{1 + a_{BC}z^{-1}}.$$
(5.66)

With $A_1(z) = -A(z)$ the signal flow chart in Fig. 5.21 shows a first-order low-pass filter and a first-order low-frequency shelving filter.

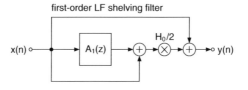

Figure 5.21 Low-frequency shelving filter and first-order low-pass filter.

The decomposition of a denormalized transfer function of a first-order high-frequency shelving filter can be given in the form

$$H(s) = \frac{s V_0 + \omega_c}{s + \omega_c}$$

$$= 1 + H_0 \frac{s}{s + \omega_c}$$
(5.67)

$$= 1 + \frac{H_0}{2}\left[1 + \frac{s - \omega_c}{s + \omega_c}\right]$$
(5.68)

where

$$V_0 = H(s = \infty),$$
(5.69)

$$H_0 = V_0 - 1.$$
(5.70)

The transfer function results by adding a high-pass filter to a constant. Applying the bilinear transformation to (5.68) gives

$$H(z) = 1 + \frac{H_0}{2}[1 + A(z)]$$
(5.71)

with

$$A(z) = -\frac{z^{-1} + a_B}{1 + a_B z^{-1}}.$$ (5.72)

For cut, the decomposition can be given by

$$H(s) = \frac{s + \omega_c}{s/V_0 + \omega_c}$$ (5.73)

$$= 1 + \underbrace{(V_0 - 1)}_{H_0} \frac{s}{s + V_0 \omega_c}$$ (5.74)

$$= 1 + \frac{H_0}{2}\left[1 + \frac{s - V_0 \omega_c}{s + V_0 \omega_c}\right],$$ (5.75)

which in turn results in (5.71) after a bilinear transformation. The boost and cut parameters can be calculated as

$$a_B = \frac{\tan(\omega_c T/2) - 1}{\tan(\omega_c T/2) + 1},$$ (5.76)

$$a_C = \frac{V_0 \tan(\omega_c T/2) - 1}{V_0 \tan(\omega_c T/2) + 1}.$$ (5.77)

The transfer function of a first-order high-frequency shelving filter can then be written as

$$H(z) = \frac{1 + (1 - a_{BC})\frac{H_0}{2} + (a_{BC} + (a_{BC} - 1)\frac{H_0}{2})z^{-1}}{1 + a_{BC} z^{-1}}.$$ (5.78)

With $A_1(z) = -A(z)$ the signal flow chart in Fig. 5.22 shows a first-order high-pass filter and a high-frequency shelving filter.

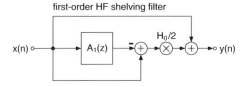

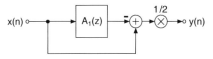

Figure 5.22 First-order high-frequency shelving and high-pass filters.

The implementation of a second-order peak filter can be carried out with a low-pass to band-pass transformation of a first-order shelving filter. But the addition of a second-order band-pass filter to a constant branch also results in a peak filter. With the help of an all-pass implementation of a band-pass filter as given by

$$H(z) = \tfrac{1}{2}[1 - A_2(z)] \tag{5.79}$$

and

$$A_2(z) = \frac{-a_B + (d - da_B)z^{-1} + z^{-2}}{1 + (d - da_B)z^{-1} - a_B z^{-2}}, \tag{5.80}$$

a second-order peak filter can be expressed as

$$H(z) = 1 + \frac{H_0}{2}[1 - A_2(z)]. \tag{5.81}$$

The bandwidth parameters a_B and a_C for boost and cut are given

$$a_B = \frac{\tan(\omega_b T/2) - 1}{\tan(\omega_b T/2) + 1}, \tag{5.82}$$

$$a_C = \frac{\tan(\omega_b T/2) - V_0}{\tan(\omega_b T/2) + V_0}. \tag{5.83}$$

The center frequency parameter d and the coefficient H_0 are given by

$$d = -\cos(\Omega_c), \tag{5.84}$$

$$V_0 = H(e^{j\Omega_c}), \tag{5.85}$$

$$H_0 = V_0 - 1. \tag{5.86}$$

The transfer function of a second-order peak filter results in

$$H(z) = \frac{1 + (1 + a_{BC})\frac{H_0}{2} + d(1 - a_{BC})z^{-1} + (-a_{BC} - (1 + a_{BC})\frac{H_0}{2})z^{-2}}{1 + d(1 - a_{BC})z^{-1} - a_{BC}z^{-2}}. \tag{5.87}$$

The signal flow charts for a second-order peak filter and a second-order band-pass filter are shown in Fig. 5.23.

second-order peak filter

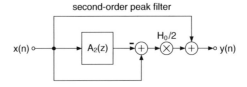

second-order band-pass filter

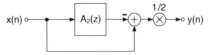

Figure 5.23 Second-order peak filter and band-pass filter.

The frequency responses for high-frequency shelving, low-frequency shelving and peak filters are shown in Figs 5.24, 5.25 and 5.26.

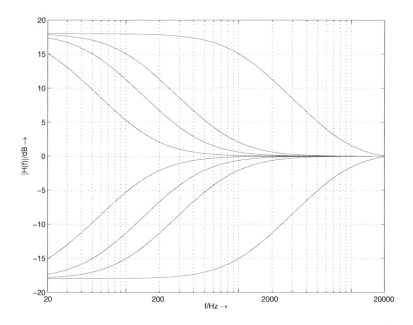

Figure 5.24 Low-frequency first-order shelving filter ($G = \pm 18$ dB, $f_c = 20$, 50, 100, 1000 Hz).

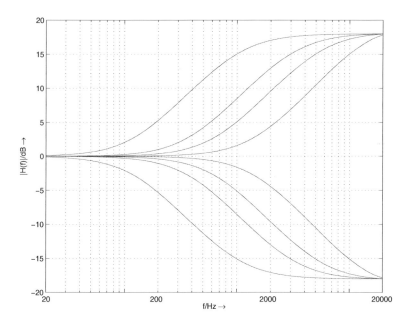

Figure 5.25 First-order high-frequency shelving filter ($G = \pm 18$ dB, $f_c = 1$, 3, 5, 10, 16 kHz).

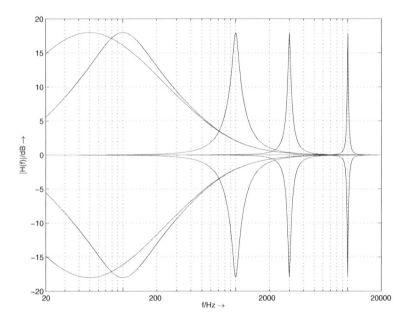

Figure 5.26 Second-order peak filter ($G = \pm 18\,\text{dB}$, $f_c = 50,\ 100,\ 1000,\ 3000,\ 10000\,\text{Hz}$, $f_b = 100\,\text{Hz}$).

5.2.3 Quantization Effects

The limited word-length for digital recursive filters leads to two different types of quantization error. The quantization of the coefficients of a digital filter results in linear distortion which can be observed as a deviation from the ideal frequency response. The quantization of the signal inside a filter structure is responsible for the maximum dynamic range and determines the noise behavior of the filter. Owing to rounding operations in a filter structure, roundoff noise is produced. Another effect of the signal quantization is limit cycles. These can be classified as overflow limit cycles, small-scale limit cycles and limit cycles correlated with the input signal. Limit cycles are very disturbing owing to their small-band (sinusoidal) nature. The overflow limit cycles can be avoided by suitable scaling of the input signal. The effects of other errors mentioned above can be reduced by increasing the word-lengths of the coefficient and the state variables of the filter structure.

The noise behavior and coefficient sensitivity of a filter structure depend on the topology and the cutoff frequency (position of the poles in the Z-domain) of the filter. Since common audio filters operate between 20 Hz and 20 kHz at a sampling rate of 48 kHz, the filter structures are subjected to specially strict criteria with respect to error behavior. The frequency range for equalizers is between 20 Hz and 4–6 kHz because the human voice and many musical instruments have their formants in that frequency region. For given coefficient and signal word-lengths (as in a digital signal processor), a filter structure with low roundoff noise for audio application can lead to a suitable solution. For this, the following second-order filter structures are compared.

The basis of the following considerations is the relationship between the coefficient sensitivity and roundoff noise. This was first stated by Fettweis [Fet72]. By increasing the pole density in a certain region of the z-plane, the coefficient sensitivity and the roundoff noise of the filter structure are reduced. Owing to these improvements, the coefficient word-length as well as signal word-length can be reduced. Work in designing digital filters with minimum word-length for coefficients and state variables was first carried out by Avenhaus [Ave71].

Typical audio filters like high-/low-pass, peak/shelving filters can be described by the second-order transfer function

$$H(z) = \frac{a_0 + a_1 z^{-1} + a_2 z^{-2}}{1 + b_1 z^{-1} + b_2 z^{-2}}. \tag{5.88}$$

The recursive part of the difference equation which can be derived from the transfer function (5.88) is considered more closely, since it plays a major role in affecting the error behavior. Owing to the quantization of the coefficients in the denominator in (5.88), the distribution of poles in the z-plane is restricted (see Fig. 5.27 for 6-bit quantization of coefficients). The pole distribution in the second quadrant of the z-plane is the mirror image of the first quadrant. Figure 5.28 shows a block diagram of the recursive part. Another equivalent representation of the denominator is given by

$$H(z) = \frac{N(z)}{1 - 2r \cos \varphi z^{-1} + r^2 z^{-2}}. \tag{5.89}$$

Here r is the radius and φ the corresponding phase of the complex poles. By quantizing these parameters, the pole distribution is altered, in contrast to the case where b_1 and b_2 are quantized as in (5.88).

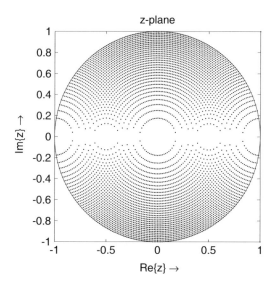

Figure 5.27 Direct-form structure – pole distribution (6-bit quantization).

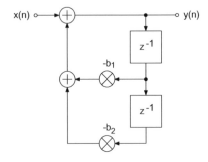

Figure 5.28 Direct-form structure – block diagram of recursive part.

The state variable structure [Mul76, Bom85] is based on the approach by Gold and Rader [Gol67], which is given by

$$H(z) = \frac{N(z)}{1 - 2\text{Re}\{z_\infty\}z^{-1} + (\text{Re}\{z_\infty\}^2 + \text{Im}\{z_\infty\}^2)z^{-2}}. \tag{5.90}$$

The possible pole locations are shown in Fig. 5.29 for 6-bit quantization (a block diagram of the recursive part is shown in Fig. 5.30). Owing to the quantization of real and imaginary parts, a uniform grid of different pole locations results. In contrast to direct quantization of the coefficients b_1 and b_2 in the denominator, the quantization of the real and imaginary parts leads to an increase in the pole density at $z = 1$. The possible pole locations in the second quadrant in the z-plane are the mirror images of the ones in the first quadrant.

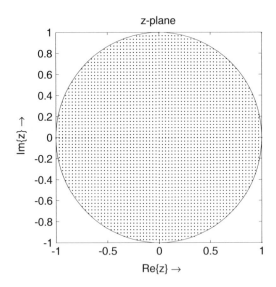

Figure 5.29 Gold and Rader – pole distribution (6-bit quantization).

In [Kin72] a filter structure is suggested which has a pole distribution as shown in Fig. 5.31 (for a block diagram of the recursive part, see Fig. 5.32).

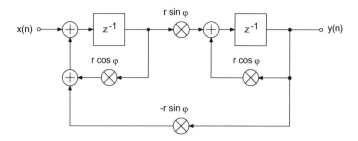

Figure 5.30 Gold and Rader – block diagram of recursive part.

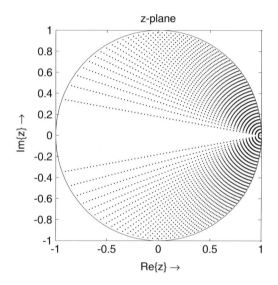

Figure 5.31 Kingsbury – pole distribution (6-bit quantization).

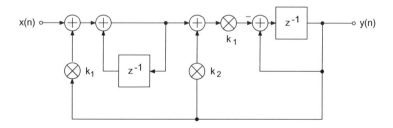

Figure 5.32 Kingsbury – block diagram of recursive part.

The corresponding transfer function,

$$H(z) = \frac{N(z)}{1 - (2 - k_1 k_2 - k_1^2)z^{-1} + (1 - k_1 k_2)z^{-2}}, \tag{5.91}$$

shows that in this case the coefficients b_1 and b_2 can be obtained by a linear combination of the quantized coefficients k_1 and k_2. The distance d of the pole from the point $z = 1$ determines the coefficients

$$k_1 = d = \sqrt{1 - 2r \cos \varphi + r^2}, \tag{5.92}$$

$$k_2 = \frac{1 - r^2}{k_1}, \tag{5.93}$$

as illustrated in Fig. 5.33.

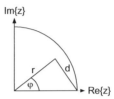

Figure 5.33 Geometric interpretation.

The filter structures under consideration showed that by a suitable linear combination of quantized coefficients, any desired pole distribution can be obtained. An increase of the pole density at $z = 1$ can be achieved by influencing the linear relationship between the coefficient k_1 and the distance d from $z = 1$ [Zöl89, Zöl90]. The nonlinear relationship of the new coefficients gives the following structure with the transfer function

$$H(z) = \frac{N(z)}{1 - (2 - z_1 z_2 - z_1^3)z^{-1} + (1 - z_1 z_2)z^{-2}} \tag{5.94}$$

and coefficients

$$z_1 = \sqrt[3]{1 + b_1 + b_2}, \tag{5.95}$$

$$z_2 = \frac{1 - b_2}{z_1}, \tag{5.96}$$

with

$$z_1 = \sqrt[3]{d^2}. \tag{5.97}$$

The pole distribution of this structure is shown in Fig. 5.34. The block diagram of the recursive part is illustrated in Fig. 5.35. The increase in the pole density at $z = 1$, in contrast to previous pole distributions is observed. The pole distributions of the Kingsbury and Zölzer structures show a decrease in the pole density for higher frequencies. For the pole density, a symmetry with respect to the imaginary axis as in the case of the direct-form structure and the Gold and Rader structure is not possible. But changing the sign in the recursive part of the difference equation results in a mirror image of the pole density. The mirror image can be achieved through a change of sign in the denominator polynomial. The denominator polynomial

$$D(z) = 1 \overset{!}{\pm} (2 - z_1 z_2 - z_1^3)z^{-1} + (1 - z_1 z_2)z^{-2} \tag{5.98}$$

shows that the real part depends on the coefficient of z^{-1}.

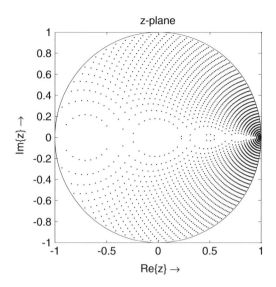

Figure 5.34 Zölzer – pole distribution (6-bit quantization).

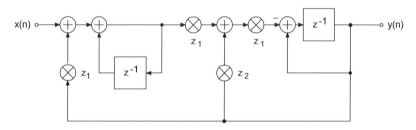

Figure 5.35 Zölzer – block diagram of recursive part.

Analytical Comparison of Noise Behavior of Different Filter Structures

In this section, recursive filter structures are analyzed in terms of their noise behavior in fixed-point arithmetic [Zöl89, Zöl90, Zöl94]. The block diagrams provide the basis for an analytical calculation of noise power owing to the quantization of state variables. First of all, the general case is considered in which quantization is performed after multiplication. For this purpose, the transfer function $G_i(z)$ of every multiplier output to the output of the filter structure is determined.

For this error analysis it is assumed that the signal within the filter structure covers the whole dynamic range so that the quantization error $e_i(n)$ is not correlated with the signal. Consecutive quantization error samples are not correlated with each other so that a uniform power density spectrum results [Sri77]. It can also be assumed that different quantization errors $e_i(n)$ are uncorrelated within the filter structure. Owing to the uniform distribution of the quantization error, the variance can be given by

$$\sigma_E^2 = \frac{Q^2}{12}. \tag{5.99}$$

The quantization error is added at every point of quantization and is filtered by the corresponding transfer function $G(z)$ to the output of the filter. The variance of the output quantization noise (due to the noise source $e(n)$) is given by

$$\sigma_{ye}^2 = \sigma_E^2 \frac{1}{2\pi j} \oint_{z=e^{j\Omega}} G(z)G(z^{-1})z^{-1}\,dz.\qquad(5.100)$$

Exact solutions for the ring integral (5.100) can be found in [Jur64] for transfer functions up to the fourth order. With the L_2 norm of a periodic function

$$\|G\|_2 = \left[\frac{1}{2\pi}\int_{-\pi}^{\pi}|G(e^{j\Omega})|^2\,d\Omega\right]^{\frac{1}{2}},\qquad(5.101)$$

the superposition of the noise variances leads to the total output noise variance

$$\sigma_{ye}^2 = \sigma_E^2 \sum_i \|G_i\|_2^2.\qquad(5.102)$$

The signal-to-noise ratio for a full-range sinusoid can be written as

$$\text{SNR} = 10\log_{10}\frac{0.5}{\sigma_{ye}^2}\ \text{dB}.\qquad(5.103)$$

The ring integral

$$I_n = \frac{1}{2\pi j}\oint_{z=e^{j\Omega}} \frac{A(z)A(z^{-1})}{B(z)B(z^{-1})}z^{-1}\,dz\qquad(5.104)$$

is given in [Jur64] for first-order systems by

$$G(z) = \frac{a_0 z + a_1}{b_0 z + b_1},\qquad(5.105)$$

$$I_1 = \frac{(a_0^2 + a_1^2)b_0 - 2a_0 a_1 b_1}{b_0(b_0^2 - b_1^2)},\qquad(5.106)$$

and for second-order systems by

$$G(z) = \frac{a_0 z^2 + a_1 z + a_2}{b_0 z^2 + b_1 z + b_2},\qquad(5.107)$$

$$I_2 = \frac{A_0 b_0 c_1 - A_1 b_0 b_1 + A_2(b_1^2 - b_2 c_1)}{b_0[(b_0^2 - b_2^2)c_1 - (b_0 b_1 - b_1 b_2)b_1]},\qquad(5.108)$$

$$A_0 = a_0^2 + a_1^2 + a_2^2,\qquad(5.109)$$

$$A_1 = 2(a_0 a_1 + a_1 a_2),\qquad(5.110)$$

$$A_2 = 2a_0 a_2,\qquad(5.111)$$

$$c_1 = b_0 + b_2.\qquad(5.112)$$

In the following, an analysis of the noise behavior for different recursive filter structures is presented. The noise transfer functions of individual recursive parts are responsible for noise shaping.

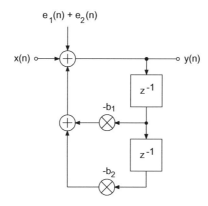

Figure 5.36 Direct form with additive error signal.

Table 5.6 Direct form – (a) noise transfer function, (b) quadratic L_2 norm and (c) output noise variance in the case of quantization after every multiplication.

$$(a) \quad G_1(z) = G_2(z) = \frac{z^2}{z^2 + b_1 z + b_2}$$

$$(b) \quad \|G_1\|_2^2 = \|G_2\|_2^2 = \frac{1 + b_2}{1 - b_2} \frac{1}{(1 + b_2)^2 - b_1^2}$$

$$(c) \quad \sigma_{ye}^2 = \sigma_E^2 2 \frac{1 + b_2}{1 - b_2} \frac{1}{(1 + b_2)^2 - b_1^2}$$

The error transfer function of a second-order direct-form structure (see Fig. 5.36) has only complex poles (see Table 5.6).

The implementation of poles near the unit circle leads to high amplification of the quantization error. The effect of the pole radius on the noise variance can be observed in the equation for output noise variance. The coefficient $b_2 = r^2$ approaches 1, which leads to a huge increase in the output noise variance.

The Gold and Rader filter structure (Fig. 5.37) has an output noise variance that depends on the pole radius (see Table 5.7) and is independent of the pole phase. The latter fact is because of the uniform grid of the pole distribution. An additional zero on the real axis ($z = r \cos \varphi$) directly beneath the poles reduces the effect of the complex poles.

The Kingsbury filter (Fig. 5.38 and Table 5.8) and the Zölzer filter (Fig. 5.39 and Table 5.9), which is derived from it, show that the noise variance depends on the pole radius. The noise transfer functions have a zero at $z = 1$ in addition to the complex poles. This zero reduces the amplifying effect of the pole near the unit circle at $z = 1$.

Figure 5.40 shows the signal-to-noise ratio versus the cutoff frequency for the four filter structures presented above. The signals are quantized to 16 bits. Here, the poles move with increasing cutoff frequency on the curve characterized by the Q-factor $Q_\infty = 0.7071$ in the z-plane. For very small cutoff frequencies, the Zölzer filter shows an improvement of 3 dB in terms of signal-to-noise ratio compared with the Kingsbury filter and an improvement of

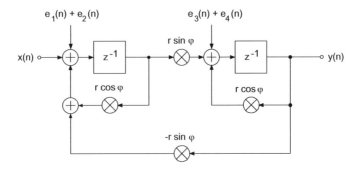

Figure 5.37 Gold and Rader structure with additive error signals.

Table 5.7 Gold and Rader – (a) noise transfer function, (b) quadratic L_2 norm and (c) output noise variance in the case of quantization after every multiplication.

(a)
$$G_1(z) = G_2(z) = \frac{r \sin \varphi}{z^2 - 2r \cos \varphi z + r^2}$$

$$G_3(z) = G_4(z) = \frac{z - r \cos \varphi}{z^2 - 2r \cos \varphi z + r^2}$$

(b)
$$\|G_1\|_2^2 = \|G_2\|_2^2 = \frac{1 + b_2}{1 - b_2} \frac{(r \sin \varphi)^2}{(1 + b_2)^2 - b_1^2}$$

$$\|G_3\|_2^2 = \|G_4\|_2^2 = \frac{1}{1 - b_2} \frac{[1 + (r \sin \varphi)^2](1 + b_2)^2 - b_1^2}{(1 + b_2)^2 - b_1^2}$$

(c)
$$\sigma_{ye}^2 = \sigma_E^2 \, 2 \, \frac{1}{1 - b_2}$$

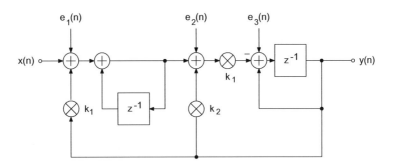

Figure 5.38 Kingsbury structure with additive error signals.

6 dB compared with the Gold and Rader filter. Up to 5 kHz, the Zölzer filter yields better results (see Fig. 5.41). From 6 kHz onwards, the reduction of pole density in this filter leads to a decrease in the signal-to-noise ratio (see Fig. 5.41).

Table 5.8 Kingsbury – (a) noise transfer function, (b) quadratic L_2 norm and (c) output noise variance in the case of quantization after every multiplication.

$$G_1(z) = \frac{-k_1 z}{z^2 - (2 - k_1 k_2 - k_1^2)z + (1 - k_1 k_2)}$$

(a) $$G_2(z) = \frac{-k_1(z - 1)}{z^2 - (2 - k_1 k_2 - k_1^2)z + (1 - k_1 k_2)}$$

$$G_3(z) = \frac{z - 1}{z^2 - (2 - k_1 k_2 - k_1^2)z + (1 - k_1 k_2)}$$

$$\|G_1\|_2^2 = \frac{1}{k_1 k_2} \frac{2 - k_1 k_2}{2(2 - k_1 k_2) - k_1^2}$$

(b) $$\|G_2\|_2^2 = \frac{k_1}{k_2} \frac{2}{2(2 - k_1 k_2) - k_1^2}$$

$$\|G_3\|_2^2 = \frac{1}{k_1 k_2} \frac{2}{2(2 - k_1 k_2) - k_1^2}$$

(c) $$\sigma_{ye}^2 = \sigma_E^2 2 \frac{5 + 2b_1 + 3b_2}{(1 - b_2)(1 + b_2 - b_1)}$$

Table 5.9 Zölzer – (a) noise transfer function, (b) quadratic L_2 norm and (c) output noise variance in the case of quantization after every multiplication.

$$G_1(z) = \frac{-z_1^2 z}{z^2 - (2 - z_1 z_2 - z_1^3)z + (1 - z_1 z_2)}$$

(a) $$G_2(z) = G_3(z) = \frac{-z_1(z - 1)}{z^2 - (2 - z_1 z_2 - z_1^3)z + (1 - z_1 z_2)}$$

$$G_4(z) = \frac{z - 1}{z^2 - (2 - z_1 z_2 - z_1^3)z + (1 - z_1 z_2)}$$

$$\|G_1\|_2^2 = \frac{z_1^4}{z_1 z_2} \frac{2 - z_1 z_2}{2z_1^3(2 - z_1 z_2) - z_1^6}$$

(b) $$\|G_2\|_2^2 = \|G_3\|_2^2 = \frac{z_1^6}{z_1 z_2} \frac{2}{2z_1^3(2 - z_1 z_2) - z_1^6}$$

$$\|G_4\|_2^2 = \frac{z_1^3}{z_1 z_2} \frac{2}{2z_1^3(2 - z_1 z_2) - z_1^6}$$

(c) $$\sigma_{ye}^2 = \sigma_E^2 2 \frac{6 + 4(b_1 + b_2) + (1 + b_2)(1 + b_1 + b_2)^{1/3}}{(1 - b_2)(1 + b_2 - b_1)}$$

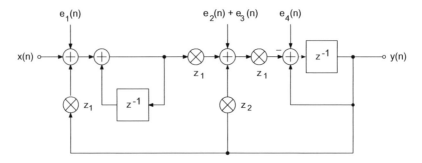

Figure 5.39 Zölzer structure with additive error signals.

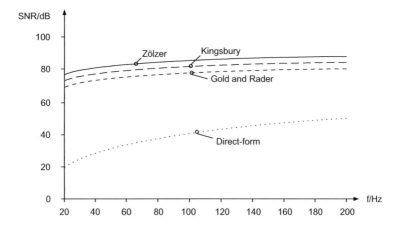

Figure 5.40 SNR vs. cutoff frequency – quantization of products ($f_c < 200\,\text{Hz}$).

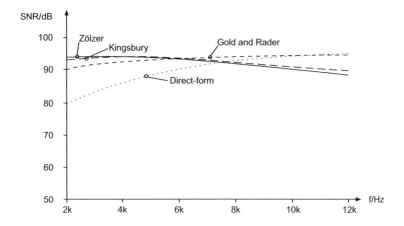

Figure 5.41 SNR vs. cutoff frequency – quantization of products ($f_c > 2\,\text{kHz}$).

With regard to the implementation of the these filters with digital signal processors, a quantization after every multiplication is not necessary. Quantization takes place when the accumulator has to be stored in memory. This can be seen in Figs 5.42–5.45 by introducing quantizers where they really occur. The resulting output noise variances are also shown. The signal-to-noise ratio is plotted versus the cutoff frequency in Figs 5.46 and 5.47. In the case of direct-form and Gold and Rader filters, the signal-to-noise ratio increases by 3 dB whereas the output noise variance for the Kingsbury filter remains unchanged. The Kingsbury filter and the Gold and Rader filters exhibit similar results up to a frequency of 200 kHz (see Fig. 5.46). The Zölzer filter demonstrates an improvement of 3 dB compared with these structures. For frequencies of up to 2 kHz (see Fig. 5.47) it is seen that the increased pole density leads to an improvement of the signal-to-noise ratio as well as a reduced effect due to coefficient quantization.

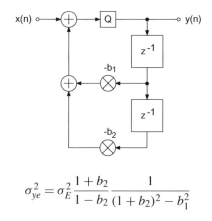

$$\sigma_{ye}^2 = \sigma_E^2 \frac{1 + b_2}{1 - b_2} \frac{1}{(1 + b_2)^2 - b_1^2}$$

Figure 5.42 Direct-form filter – quantization after accumulator.

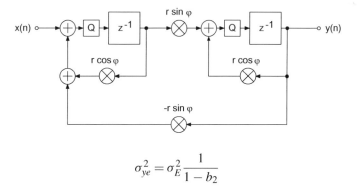

$$\sigma_{ye}^2 = \sigma_E^2 \frac{1}{1 - b_2}$$

Figure 5.43 Gold and Rader filter – quantization after accumulator.

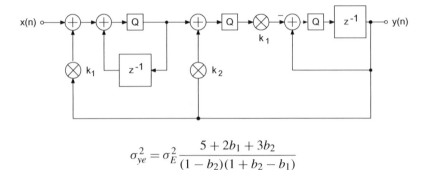

$$\sigma_{ye}^2 = \sigma_E^2 \frac{5 + 2b_1 + 3b_2}{(1 - b_2)(1 + b_2 - b_1)}$$

Figure 5.44 Kingsbury filter – quantization after accumulator.

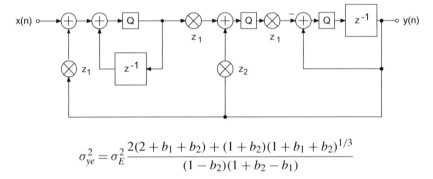

$$\sigma_{ye}^2 = \sigma_E^2 \frac{2(2 + b_1 + b_2) + (1 + b_2)(1 + b_1 + b_2)^{1/3}}{(1 - b_2)(1 + b_2 - b_1)}$$

Figure 5.45 Zölzer filter – quantization after accumulator.

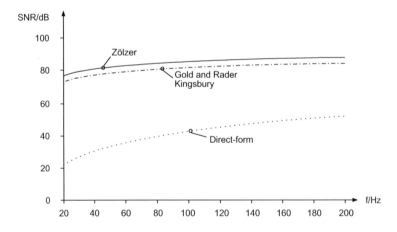

Figure 5.46 SNR vs. cutoff frequency – quantization after accumulator ($f_c < 200\,\text{Hz}$).

Noise Shaping in Recursive Filters

The analysis of the noise transfer function of different structures shows that for three structures with low roundoff noise a zero at $z = 1$ occurs in the transfer functions $G(z)$

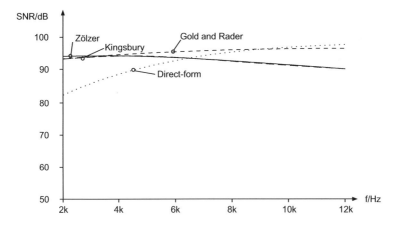

Figure 5.47 SNR vs. cutoff frequency – quantization after accumulator ($f_c > 2\,\text{kHz}$).

of the error signals in addition to the complex poles. This zero near the poles reduces the amplifying effect of the pole. If it is now possible to introduce another zero into the noise transfer function then the effect of the poles is compensated for to a larger extent. The procedure of feeding back the quantization error as shown in Chapter 2 produces an additional zero in the noise transfer function [Tra77, Cha78, Abu79, Bar82, Zöl89]. The feedback of the quantization error is first demonstrated with the help of the direct-form structure as shown in Fig. 5.48. This generates a zero at $z = 1$ in the noise transfer function given by

$$G_{1.O}(z) = \frac{1 - z^{-1}}{1 + b_1 z^{-1} + b_2 z^{-2}}. \tag{5.113}$$

The resulting variance σ^2 of the quantization error at the output of the filter is presented in Fig. 5.48. In order to produce two zeros at $z = 1$, the quantization error is fed back over two delays weighted 2 and -1 (see Fig. 5.48b). The noise transfer function is, hence, given by

$$G_{2.O}(z) = \frac{1 - 2z^{-1} + z^{-2}}{1 + b_1 z^{-1} + b_2 z^{-2}}. \tag{5.114}$$

The signal-to-noise ratio of the direct-form is plotted versus the cutoff frequency in Fig. 5.49. Even a single zero significantly improves the signal-to-noise ratio in the direct form. The coefficients b_1 and b_2 approach -2 and 1 respectively with the decrease of the cutoff frequency. With this, the error is filtered with a second-order high-pass. The introduction of the additional zeros in the noise transfer function only affects the noise signal of the filter. The input signal is only affected by the transfer function $H(z)$. If the feedback coefficients are chosen equal to the coefficients b_1 and b_2 in the denominator polynomial, complex zeros are produced that are identical to the complex poles. The noise transfer function $G(z)$ is then reduced to unity. The choice of complex zeros directly at the location of the complex poles corresponds to double-precision arithmetic.

In [Abu79] an improvement of noise behavior for the direct form in any desired location of the z-plane is achieved by placing additional simple-to-implement complex zeros near

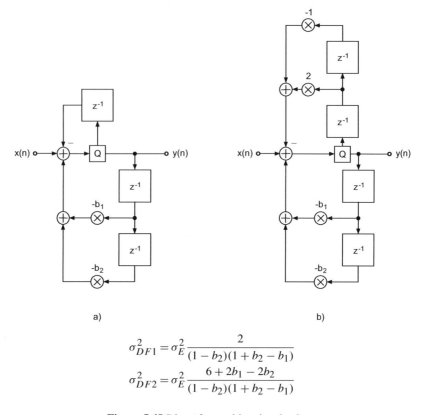

a)   b)

$$\sigma^2_{DF1} = \sigma^2_E \frac{2}{(1 - b_2)(1 + b_2 - b_1)}$$

$$\sigma^2_{DF2} = \sigma^2_E \frac{6 + 2b_1 - 2b_2}{(1 - b_2)(1 + b_2 - b_1)}$$

Figure 5.48 Direct form with noise shaping.

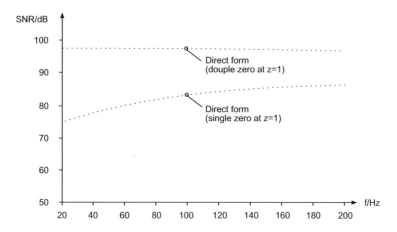

Figure 5.49 SNR – Noise shaping in direct-form filter structures.

the poles. For implementing filter algorithms with digital signal processors, these kinds of suboptimal zero are easily realized. Since the Gold and Rader, Kingsbury, and Zölzer filter structures already have zeros in their respective noise transfer functions, it is sufficient to use a simple feedback for the quantization error. By virtue of this extension, the block diagrams in Figs 5.50, 5.51 and 5.52 are obtained.

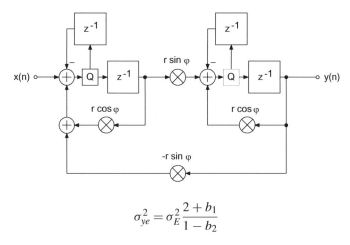

$$\sigma_{ye}^2 = \sigma_E^2 \frac{2 + b_1}{1 - b_2}$$

Figure 5.50 Gold and Rader filter with noise shaping.

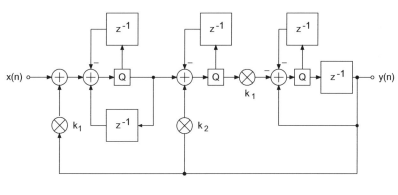

$$\sigma_{ye}^2 = \sigma_E^2 \frac{(1 + k_1^2)((1 + b_2)(6 - 2b_2) + 2b_1^2 + 8b_1) + 2k_1^2(1 + b_1 + b_2)}{(1 - b_2)(1 + b_2 - b_1)(1 + b_2 - b_1)}$$

Figure 5.51 Kingsbury filter with noise shaping.

The effect of noise shaping on signal-to-noise ratio is shown in Figs 5.53 and 5.54. The almost ideal noise behavior of all filter structures for 16-bit quantization and very small cutoff frequencies can be observed. The effect of this noise shaping for increasing cutoff frequencies is shown in Fig. 5.54. The compensating effect of the two zeros at $z = 1$ is reduced.

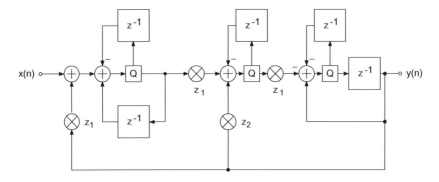

$$\sigma_{ye}^2 = \sigma_E^2 \frac{(1+z_1^2)((1+b_2)(6-2b_2) + 2b_1^2 + 8b_1) + 2z_1^4(1+b_1+b_2)}{(1-b_2)(1+b_2-b_1)(1+b_2-b_1)}$$

Figure 5.52 Zölzer filter with noise shaping.

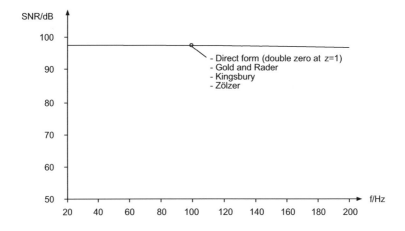

Figure 5.53 SNR – noise shaping (20–200 Hz).

Scaling

In a fixed-point implementation of a digital filter, a transfer function from the input of the filter to a junction within the filter has to be determined, as well as the transfer function from the input to the output. By scaling the input signal, it has to be guaranteed that the signals remain within the number range at each junction and at the output.

In order to calculate scaling coefficients, different criteria can be used. The L_p norm is defined as

$$L_p = \|H\|_p = \left[\frac{1}{2\pi} \int_{-\pi}^{\pi} |H(e^{j\Omega})|^p \, d\Omega \right]^{1/p}, \tag{5.115}$$

and an expression for the L_∞ norm follows for $p = \infty$:

$$L_\infty = \|H(e^{j\Omega})\|_\infty = \max_{0 \leq \Omega \leq \pi} |H(e^{j\Omega})|. \tag{5.116}$$

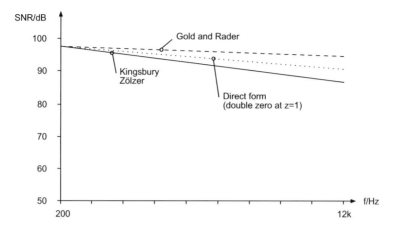

Figure 5.54 SNR – noise shaping (200–12000 Hz).

The L_∞ norm represents the maximum of the amplitude frequency response. In general, the modulus of the output is

$$|y(n)| \leq \|H\|_p \|X\|_q \tag{5.117}$$

with

$$\frac{1}{p} + \frac{1}{q} = 1, \quad p, q \geq 1. \tag{5.118}$$

For the L_1, L_2 and L_∞ norms the explanations in Table 5.10 can be used.

Table 5.10 Commonly used scaling.

p	q	
1	∞	Given max. value of input spectrum scaling w.r.t. the L_1 norm of $H(e^{j\Omega})$
∞	1	Given L_1 norm of input spectrum $X(e^{j\Omega})$ scaling w.r.t. the L_∞ norm of $H(e^{j\Omega})$
2	2	Given L_2 norm of input spectrum $X(e^{j\Omega})$ scaling w.r.t. the L_2 norm of $H(e^{j\Omega})$

With

$$|y_i(n)| \leq \|H_i(e^{j\Omega})\|_\infty \|X(e^{j\Omega})\|_1, \tag{5.119}$$

the L_∞ norm is given by

$$L_\infty = \|h_i\|_\infty = \max_{k=0}^{\infty} |h_i(k)|. \tag{5.120}$$

For a sinusoidal input signal of amplitude 1 we get $\|X(e^{j\Omega})\|_1 = 1$. For $|y_i(n)| \leq 1$ to be valid, the scaling factor must be chosen as

$$S_i = \frac{1}{\|H_i(e^{j\Omega})\|_\infty}. \tag{5.121}$$

The scaling of the input signal is carried out with the maximum of the amplitude frequency response with the goal that for $|x(n)| \leq 1$, $|y_i(n)| \leq 1$. As a scaling coefficient for the input signal the highest scaling factor S_i is chosen. To determine the maximum of the transfer function

$$\|H(e^{j\Omega})\|_\infty = \max_{0 \leq \Omega \leq \pi} |H(e^{j\Omega})| \tag{5.122}$$

of a second-order system

$$H(z) = \frac{a_0 + a_1 z^{-1} + a_2 z^{-1}}{1 + b_1 z^{-1} + b_2 z^{-1}} = \frac{a_0 z^2 + a_1 z + a_2}{z^2 + b_1 z + b_2},$$

the maximum value can be calculated as

$$|H(e^{j\Omega})|^2 = \frac{\overbrace{\frac{a_0 a_2}{b_2}}^{\alpha_0} \cos^2(\Omega) + \overbrace{\frac{a_1(a_0 + a_2)}{2b_2}}^{\alpha_1} \cos(\Omega) + \overbrace{\frac{(a_0 - a_2)^2 + a_1^2}{4b_2}}^{\alpha_2}}{\cos^2(\Omega) + \underbrace{\frac{b_1(1 + b_2)}{2b_2}}_{\beta_1} \cos(\Omega) + \underbrace{\frac{(1 - b_2)^2 + b_1^2}{4b_2}}_{\beta_2}} = S^2. \tag{5.123}$$

With $x = \cos(\Omega)$ it follows that

$$(S^2 - \alpha_0)x^2 + (\beta_1 S^2 - \alpha_1)x + (\beta_2 S^2 - \alpha_2) = 0. \tag{5.124}$$

The solution of (5.124) leads to $x = \cos(\Omega_{max / min})$ which must be real $(-1 \leq x \leq 1)$ for the maximum/minimum to occur at a real frequency. For a single solution (repeated roots) of the above quadratic equation, the discriminant must be $D = (p/2)^2 - q = 0$ $(x^2 + px + q = 0)$. It follows that

$$D = \frac{(\beta_1 S^2 - \alpha_1)^2}{4(S^2 - \alpha_0)^2} - \frac{\beta_2 S^2 - \alpha_2}{S^2 - \alpha_0} = 0 \tag{5.125}$$

and

$$S^4(\beta_1^2 - 4\beta_2) + S^2(4\alpha_2 + 4\alpha_0\beta_2 - 2\alpha_1\beta_1) + (\alpha_1^2 - 4\alpha_0\alpha_2) = 0. \tag{5.126}$$

The solution of (5.126) gives two solutions for S^2. The solution with the larger value is chosen. If the discriminant D is not greater than zero, the maximum lies at $x = 1$ $(z = 1)$ or $x = -1$ $(z = -1)$ as given by

$$S^2 = \frac{\alpha_0 + \alpha_1 + \alpha_2}{1 + \beta_1 + \beta_2} \tag{5.127}$$

or

$$S^2 = \frac{\alpha_0 - \alpha_1 + \alpha_2}{1 - \beta_1 + \beta_2}. \tag{5.128}$$

Limit Cycles

Limit cycles are periodic processes in a filter which can be measured as sinusoidal signals. They arise owing to the quantization of state variables. The different types of limit cycle and the methods necessary to prevent them are briefly listed below:

- overflow limit cycles
  - → saturation curve
  - → scaling
- limit cycles for vanishing input
  - → noise shaping
  - → dithering
- limit cycles correlated with the input signal
  - → noise shaping
  - → dithering.

5.3 Nonrecursive Audio Filters

To implement linear phase audio filters, nonrecursive filters are used. The basis of an efficient implementation is the *fast convolution*

$$y(n) = x(n) * h(n) \circ\!\!-\!\!\bullet\ Y(k) = X(k) \cdot H(k), \tag{5.129}$$

where the convolution in the time domain is performed by transforming the signal and the impulse response into the frequency domain, multiplying the corresponding Fourier transforms and inverse Fourier transform of the product into the time domain signal (see Fig. 5.55). The transform is carried out by a discrete Fourier transform of length N, such that $N = N_1 + N_2 - 1$ is valid and time-domain aliasing is avoided. First we discuss the basics. We then introduce the convolution of long sequences followed by a filter design for linear phase filters.

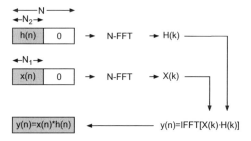

Figure 5.55 Fast convolution of signal $x(n)$ of length N_1 and impulse response $h(n)$ of length N_2 delivers the convolution result $y(n) = x(n) * h(n)$ of length $N_1 + N_2 - 1$.

5.3.1   Basics of Fast Convolution

IDFT Implementation with DFT Algorithm. The discrete Fourier transformation (DFT) is described by

$$X(k) = \sum_{n=0}^{N-1} x(n) W_N^{nk} = \mathrm{DFT}_k[x(n)], \tag{5.130}$$

$$W_N = e^{-j2\pi/N}, \tag{5.131}$$

and the inverse discrete Fourier transformation (IDFT) by

$$x(n) = \frac{1}{N} \sum_{k=0}^{N-1} X(k) W_N^{-nk}. \tag{5.132}$$

Suppressing the scaling factor $1/N$, we write

$$x'(n) = \sum_{k=0}^{N-1} X(k) W_N^{-nk} = \mathrm{IDFT}_n[X(k)], \tag{5.133}$$

so that the following symmetrical transformation algorithms hold:

$$X'(k) = \frac{1}{\sqrt{N}} \sum_{n=0}^{N-1} x(n) W_N^{nk}, \tag{5.134}$$

$$x(n) = \frac{1}{\sqrt{N}} \sum_{k=0}^{N-1} X'(k) W_N^{-nk}. \tag{5.135}$$

The IDFT differs from the DFT only by the sign in the exponential term.

An alternative approach for calculating the IDFT with the help of a DFT is described as follows [Cad87, Duh88]. We will make use of the relationships

$$x(n) = a(n) + j \cdot b(n), \tag{5.136}$$
$$j \cdot x^*(n) = b(n) + j \cdot a(n). \tag{5.137}$$

Conjugating (5.133) gives

$$x'^*(n) = \sum_{k=0}^{N-1} X^*(k) W_N^{nk}. \tag{5.138}$$

The multiplication of (5.138) by j leads to

$$j \cdot x'^*(n) = \sum_{k=0}^{N-1} j \cdot X^*(k) W_N^{nk}. \tag{5.139}$$

Conjugating and multiplying (5.139) by j results in

$$x'(n) = j \cdot \left[\sum_{k=0}^{N-1} (j \cdot X^*(k) W_N^{nk} \right]^*. \tag{5.140}$$

An interpretation of (5.137) and (5.140) suggests the following way of performing the IDFT with the DFT algorithm:

1. Exchange the real with the imaginary part of the spectral sequence

$$Y(k) = Y_I(k) + jY_R(k).$$

2. Transform with DFT algorithm

$$\text{DFT}[Y(k)] = y_I(n) + jy_R(n).$$

3. Exchange the real with the imaginary part of the time sequence

$$y(n) = y_R(n) + jy_I(n).$$

For implementation on a digital signal processor, the use of DFT saves memory for IDFT.

Discrete Fourier Transformation of Two Real Sequences. In many applications, stereo signals that consist of a left and right channel are processed. With the help of the DFT, both channels can be transformed simultaneously into the frequency domain [Sor87, Ell82].

For a real sequence $x(n)$ we obtain

$$X(k) = X^*(-k), \quad k = 0, 1, \ldots, N-1 \tag{5.141}$$
$$= X^*(N-k). \tag{5.142}$$

For a discrete Fourier transformation of two real sequences $x(n)$ and $y(n)$, a complex sequence is first formed according to

$$z(n) = x(n) + jy(n). \tag{5.143}$$

The Fourier transformation gives

$$
\begin{aligned}
\text{DFT}[z(n)] &= \text{DFT}[x(n) + jy(n)] \\
&= Z_R(k) + jZ_I(k) \tag{5.144} \\
&= Z(k), \tag{5.145}
\end{aligned}
$$

where

$$
\begin{aligned}
Z(k) &= Z_R(k) + jZ_I(k) \tag{5.146} \\
&= X_R(k) + jX_I(k) + j[Y_R(k) + jY_I(k)] \tag{5.147} \\
&= X_R(k) - Y_I(k) + j[X_I(k) + Y_R(k)]. \tag{5.148}
\end{aligned}
$$

Since $x(n)$ and $y(n)$ are real sequences, it follows from (5.142) that

$$
\begin{aligned}
Z(N-k) &= Z_R(N-k) + jZ_I(N-k) = Z^*(k) \tag{5.149} \\
&= X_R(k) - jX_I(k) + j[Y_R(k) - jY_I(k)] \tag{5.150} \\
&= X_R(k) + Y_I(k) - j[X_I(k) - Y_R(k)]. \tag{5.151}
\end{aligned}
$$

Considering the real part of $Z(k)$, adding (5.148) and (5.151) gives

$$
\begin{aligned}
2X_R(k) &= Z_R(k) + Z_R(N-k) \tag{5.152} \\
&\rightarrow X_R(k) = \tfrac{1}{2}[Z_R(k) + Z_R(N-k)], \tag{5.153}
\end{aligned}
$$

and subtraction of (5.151) from (5.148) results in

$$2Y_I(k) = Z_R(N - k) - Z_R(k) \tag{5.154}$$
$$\to Y_I(k) = \tfrac{1}{2}[Z_R(N - k) - Z_R(k)]. \tag{5.155}$$

Considering the imaginary part of $Z(k)$, adding (5.148) and (5.151) gives

$$2Y_R(k) = Z_I(k) + Z_I(N - k) \tag{5.156}$$
$$\to Y_R(k) = \tfrac{1}{2}[Z_I(k) + Z_I(N - k)], \tag{5.157}$$

and subtraction of (5.151) from (5.148) results in

$$2X_I(k) = Z_I(k) - Z_I(N - k) \tag{5.158}$$
$$\to X_I(k) = \tfrac{1}{2}[Z_I(k) - Z_I(N - k)]. \tag{5.159}$$

Hence, the spectral functions are given by

$$X(k) = \mathrm{DFT}[x(n)] = X_R(k) + jX_I(k) \tag{5.160}$$
$$= \frac{1}{2}[Z_R(k) + Z_R(N - k)]$$
$$+ j\frac{1}{2}[Z_I(k) - Z_I(N - k)], \quad k = 0, 1, \ldots, \frac{N}{2}, \tag{5.161}$$
$$Y(k) = \mathrm{DFT}[y(n)] = Y_R(k) + jY_R(k) \tag{5.162}$$
$$= \frac{1}{2}[Z_I(k) + Z_I(N - k)]$$
$$+ j\frac{1}{2}[Z_R(N - k) - Z_R(k)], \quad k = 0, 1, \ldots, \frac{N}{2}, \tag{5.163}$$

and

$$X_R(k) + jX_I(k) = X_R(N - k) - jX_I(N - k) \tag{5.164}$$
$$Y_R(k) + jY_I(k) = Y_R(N - k) - jY_I(N - k), \quad k = \frac{N}{2} + 1, \ldots, N - 1. \tag{5.165}$$

Fast Convolution if Spectral Functions are Known. The spectral functions $X(k)$, $Y(k)$ and $H(k)$ are known. With the help of (5.148), the spectral sequence can be formed by

$$Z(k) = Z_R(k) + jZ_I(k) \tag{5.166}$$
$$= X_R(k) - Y_I(k) + j[X_I(k) + Y_R(k)], \quad k = 0, 1, \ldots, N - 1. \tag{5.167}$$

Filtering is done by multiplication in the frequency domain:

$$Z'(k) = [Z_R(k) + jZ_I(k)][H_R(k) + jH_I(k)]$$
$$= Z_R(k)H_R(k) - Z_I(k)H_I(k) + j[Z_R(k)H_I(k) + Z_I(k)H_R(k)]. \tag{5.168}$$

The inverse transformation gives

$$z'(n) = [x(n) + jy(n)] * h(n) = x(n) * h(n) + jy(n) * h(n) \tag{5.169}$$
$$= \mathrm{IDFT}[Z'(k)]$$
$$= z'_R(n) + jz'_I(n), \tag{5.170}$$

so that the filtered output sequence is given by

$$x'(n) = z'_R(n), \qquad (5.171)$$
$$y'(n) = z'_I(n). \qquad (5.172)$$

The filtering of a stereo signal can hence be done by transformation into the frequency domain, multiplication of the spectral functions and inverse transformation of left and right channels.

5.3.2 Fast Convolution of Long Sequences

The fast convolution of two real input sequences $x_l(n)$ and $x_{l+1}(n)$ of length N_1 with the impulse response $h(n)$ of length N_2 leads to the output sequences

$$y_l(n) = x_l(n) * h(n), \qquad (5.173)$$
$$y_{l+1}(n) = x_{l+1}(n) * h(n), \qquad (5.174)$$

of length $N_1 + N_2 - 1$. The implementation of a nonrecursive filter with fast convolution becomes more efficient than the direct implementation of an FIR filter for filter lengths $N > 30$. Therefore the following procedure will be performed:

- Formation of a complex sequence

$$z(n) = x_l(n) + jx_{l+1}(n). \qquad (5.175)$$

- Fourier transformation of the impulse response $h(n)$ that is padded with zeros to a length $N \geq N_1 + N_2 - 1$,

$$H(k) = \text{DFT}[h(n)] \quad \text{(FFT-length } N\text{)}. \qquad (5.176)$$

- Fourier transformation of the sequence $z(n)$ that is padded with zeros to a length $N \geq N_1 + N_2 - 1$,

$$Z(k) = \text{DFT}[z(n)] \quad \text{(FFT-length } N\text{)}. \qquad (5.177)$$

- Formation of a complex output sequence

$$e(n) = \text{IDFT}[Z(k)H(k)] \qquad (5.178)$$
$$= z(n) * h(n) \qquad (5.179)$$
$$= x_l(n) * h(n) + jx_{l+1}(n) * h(n). \qquad (5.180)$$

- Formation of a real output sequence

$$y_l(n) = \text{Re}\{e(n)\}, \qquad (5.181)$$
$$y_{l+1}(n) = \text{Im}\{e(n)\}. \qquad (5.182)$$

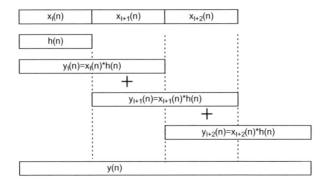

Figure 5.56 Fast convolution with partitioning of the input signal $x(n)$ into blocks of length L.

For the convolution of an infinite-length input sequence (see Fig. 5.56) with an impulse response $h(n)$, the input sequence is partitioned into sequences $x_m(n)$ of length L:

$$x_m(n) = \begin{cases} x(n), & (m-1)L \leq n \leq mL - 1, \\ 0, & \text{otherwise.} \end{cases} \tag{5.183}$$

The input sequence is given by superposition of finite-length sequences according to

$$x(n) = \sum_{m=1}^{\infty} x_m(n). \tag{5.184}$$

The convolution of the input sequence with the impulse response $h(n)$ of length M gives

$$y(n) = \sum_{k=0}^{M-1} h(k)x(n-k) \tag{5.185}$$

$$= \sum_{k=0}^{M-1} h(k) \sum_{m=1}^{\infty} x_m(n-k) \tag{5.186}$$

$$= \sum_{m=1}^{\infty} \left[\sum_{k=0}^{M-1} h(k)x_m(n-k) \right]. \tag{5.187}$$

The term in brackets corresponds to the convolution of a finite-length sequence $x_m(n)$ of length L with the impulse response of length M. The output signal can be given as superposition of convolution products of length $L + M - 1$. With these partial convolution products

$$y_m(n) = \begin{cases} \sum_{k=0}^{M-1} h(k)x_m(n-k), & (m-1)L \leq n \leq mL + M - 2, \\ 0, & \text{otherwise,} \end{cases} \tag{5.188}$$

the output signal can be written as

$$y(n) = \sum_{m=1}^{\infty} y_m(n). \tag{5.189}$$

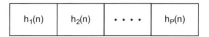

Figure 5.57 Partitioning of the impulse response $h(n)$.

If the length M of the impulse response is very long, it can be similarly partitioned into P parts each of length M/P (see Fig. 5.57). With

$$h_p\left(n - (p-1)\frac{M}{P}\right) = \begin{cases} h(n), & (p-1)\frac{M}{P} \le n \le p\frac{M}{P} - 1, \\ 0, & \text{otherwise,} \end{cases} \tag{5.190}$$

it follows that

$$h(n) = \sum_{p=1}^{P} h_p\left(n - (p-1)\frac{M}{P}\right). \tag{5.191}$$

With $M_p = pM/P$ and (5.189) the following partitioning can be done:

$$y(n) = \sum_{m=1}^{\infty}\underbrace{\left[\sum_{k=0}^{M-1} h(k)x_m(n-k)\right]}_{y_m(n)} \tag{5.192}$$

$$= \sum_{m=1}^{\infty}\left[\sum_{k=0}^{M_1-1} h(k)x_m(n-k) + \sum_{k=M_1}^{M_2-1} h(k)x_m(n-k) + \cdots + \sum_{k=M_{P-1}}^{M-1} h(k)x_m(n-k)\right]. \tag{5.193}$$

This can be rewritten as

$$y(n) = \sum_{m=1}^{\infty}\left[\underbrace{\sum_{k=0}^{M_1-1} h_1(k)x_m(n-k)}_{y_{m1}} + \underbrace{\sum_{k=0}^{M_1-1} h_2(k)x_m(n-M_1-k)}_{y_{m2}}\right.$$

$$\left.+ \underbrace{\sum_{k=0}^{M_1-1} h_3(k)x_m(n-2M_1-k)}_{y_{m3}} + \cdots + \underbrace{\sum_{k=0}^{M_1-1} h_P(k)x_m(n-(P-1)M_1-k)}_{y_{mP}}\right]$$

$$= \sum_{m=1}^{\infty}\underbrace{[y_{m1}(n) + y_{m2}(n-M_1) + \cdots + y_{mP}(n-(P-1)M_1)]}_{y_m(n)}. \tag{5.194}$$

An example of partitioning the impulse response into $P = 4$ parts is graphically shown in Fig. 5.58. This leads to

$$y(n) = \sum_{m=1}^{\infty} \Big[\underbrace{\sum_{k=0}^{M_1-1} h_1(k)x_m(n-k)}_{y_{m1}} + \underbrace{\sum_{k=0}^{M_1-1} h_2(k)x_m(n-M_1-k)}_{y_{m2}}$$

$$+ \underbrace{\sum_{k=0}^{M_1-1} h_3(k)x_m(n-2M_1-k)}_{y_{m3}} + \underbrace{\sum_{k=0}^{M_1-1} h_4(k)x_m(n-3M_1-k)}_{y_{m4}} \Big]$$

$$= \sum_{m=1}^{\infty} \underbrace{[y_{m1}(n) + y_{m2}(n-M_1) + y_{m3}(n-2M_1) + y_{m4}(n-3M_1)]}_{y_m(n)}. \quad (5.195)$$

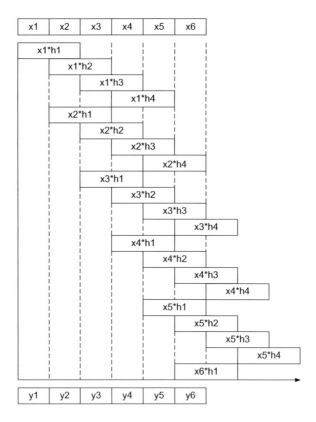

Figure 5.58 Scheme for a fast convolution with $P = 4$.

The procedure of a fast convolution by partitioning the input sequence $x(n)$ as well as the impulse response $h(n)$ is given in the following for the example in Fig. 5.58.

1. Decomposition of the impulse response $h(n)$ of length $4M$:

$$h_1(n) = h(n) \quad 0 \leq n \leq M - 1, \tag{5.196}$$
$$h_2(n - M) = h(n) \quad M \leq n \leq 2M - 1, \tag{5.197}$$
$$h_3(n - 2M) = h(n) \quad 2M \leq n \leq 3M - 1, \tag{5.198}$$
$$h_4(n - 3M) = h(n) \quad 3M \leq n \leq 4M - 1. \tag{5.199}$$

2. Zero-padding of partial impulse responses up to a length $2M$:

$$h_1(n) = \begin{cases} h_1(n), & 0 \leq n \leq M - 1, \\ 0, & M \leq n \leq 2M - 1, \end{cases} \tag{5.200}$$

$$h_2(n) = \begin{cases} h_2(n), & 0 \leq n \leq M - 1, \\ 0, & M \leq n \leq 2M - 1, \end{cases} \tag{5.201}$$

$$h_3(n) = \begin{cases} h_3(n), & 0 \leq n \leq M - 1, \\ 0, & M \leq n \leq 2M - 1, \end{cases} \tag{5.202}$$

$$h_4(n) = \begin{cases} h_4(n), & 0 \leq n \leq M - 1, \\ 0, & M \leq n \leq 2M - 1. \end{cases} \tag{5.203}$$

3. Calculating and storing

$$H_i(k) = \text{DFT}[h_i(n)], \quad i = 1, \ldots, 4 \quad \text{(FFT-length } 2M\text{)}. \tag{5.204}$$

4. Decomposition of the input sequence $x(n)$ into partial sequences $x_l(n)$ of length M:

$$x_l(n) = x(n), \quad (l - 1)M \leq n \leq lM - 1, \quad l = 1, \ldots, \infty. \tag{5.205}$$

5. Nesting partial sequences:

$$z_m(n) = x_l(n) + jx_{l+1}(n), \quad m = 1, \ldots, \infty. \tag{5.206}$$

6. Zero-padding of complex sequence $z_m(n)$ up to a length $2M$:

$$z_m(n) = \begin{cases} z_m(n), & (l - 1)M \leq n \leq lM - 1, \\ 0, & lM \leq n \leq (l + 1)M - 1. \end{cases} \tag{5.207}$$

7. Fourier transformation of the complex sequences $z_m(n)$:

$$Z_m(k) = \text{DFT}[z_m(n)] = Z_{mR}(k) + jZ_{mI}(k) \quad \text{(FFT-length } 2M\text{)}. \tag{5.208}$$

8. Multiplication in the frequency domain:

$$[Z_R(k) + jZ_I(k)][H_R(k) + jH_I(k)] = Z_R(k)H_R(k) - Z_I(k)H_I(k)$$
$$+ j[Z_R(k)H_I(k) + Z_I(k)H_R(k)], \tag{5.209}$$

$$E_{m1}(k) = Z_m(k)H_1(k) \quad k = 0, 1, \ldots, 2M - 1, \tag{5.210}$$
$$E_{m2}(k) = Z_m(k)H_2(k) \quad k = 0, 1, \ldots, 2M - 1, \tag{5.211}$$
$$E_{m3}(k) = Z_m(k)H_3(k) \quad k = 0, 1, \ldots, 2M - 1, \tag{5.212}$$
$$E_{m4}(k) = Z_m(k)H_4(k) \quad k = 0, 1, \ldots, 2M - 1. \tag{5.213}$$

9. Inverse transformation:

$$e_{m1}(n) = \text{IDFT}[Z_m(k)H_1(k)] \quad n = 0, 1, \ldots, 2M - 1, \qquad (5.214)$$
$$e_{m2}(n) = \text{IDFT}[Z_m(k)H_2(k)] \quad n = 0, 1, \ldots, 2M - 1, \qquad (5.215)$$
$$e_{m3}(n) = \text{IDFT}[Z_m(k)H_3(k)] \quad n = 0, 1, \ldots, 2M - 1, \qquad (5.216)$$
$$e_{m4}(n) = \text{IDFT}[Z_m(k)H_4(k)] \quad n = 0, 1, \ldots, 2M - 1. \qquad (5.217)$$

10. Determination of partial convolutions:

$$\text{Re}\{e_{m1}(n)\} = x_l * h_1, \qquad (5.218)$$
$$\text{Im}\{e_{m1}(n)\} = x_{l+1} * h_1, \qquad (5.219)$$
$$\text{Re}\{e_{m2}(n)\} = x_l * h_2, \qquad (5.220)$$
$$\text{Im}\{e_{m2}(n)\} = x_{l+1} * h_2, \qquad (5.221)$$
$$\text{Re}\{e_{m3}(n)\} = x_l * h_3, \qquad (5.222)$$
$$\text{Im}\{e_{m3}(n)\} = x_{l+1} * h_3, \qquad (5.223)$$
$$\text{Re}\{e_{m4}(n)\} = x_l * h_4, \qquad (5.224)$$
$$\text{Im}\{e_{m4}(n)\} = x_{l+1} * h_4. \qquad (5.225)$$

11. *Overlap-add* of partial sequences, increment $l = l + 2$ and $m = m + 1$, and back to step 5.

Based on the partitioning of the input signal and the impulse response and the following Fourier transform, the result of each single convolution is only available after a delay of one block of samples. Different methods to reduce computational complexity or overcome the block delay have been proposed [Soo90, Gar95, Ege96, Mül99, Mül01, Garc02]. These methods make use of a hybrid approach where the first part of the impulse response is used for time-domain convolution and the other parts are used for fast convolution in the frequency domain. Figure 5.59a,b demonstrates a simple derivation of the hybrid convolution scheme, which can be described by the decomposition of the transfer function as

$$H(z) = \sum_{i=0}^{M-1} z^{-iN} H_i(z), \qquad (5.226)$$

where the impulse response has length $M \cdot N$ and M is the number of smaller partitions of length N. Figure 5.59c,d shows two different signal flow graphs for the decomposition given by (5.226) of the entire transfer function. In particular, Fig. 5.59d highlights (with gray background) that in each branch $i = 1, \ldots, M - 1$ a delay of $i \cdot N$ occurs and each filter $H_i(z)$ has the same length and makes use of the same state variables. This means that they can be computed in parallel in the frequency domain with $2N$-FFTs/IFFTs and the outputs have to be delayed according to $(i - 1) \cdot N$, as shown in Fig. 5.59e. A further simplification shown in Fig. 5.59f leads to one input $2N$-FFT and block delays z^{-1} for the frequency vectors. Then, parallel multiplications with $H_i(k)$ of length $2N$ and the summation of all intermediate products are performed before one output $2N$-IFFT for the overlap-add operation in the time domain is used. The first part of the impulse response represented by $H_0(z)$ is performed by direct convolution in the time domain. The frequency and time domain parts are then overlapped and added. An alternative realization for fast convolution is based on the overlap and save operation.

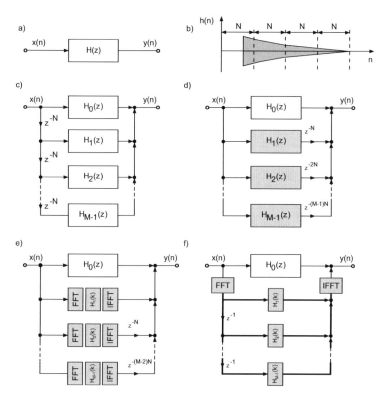

Figure 5.59 Hybrid fast convolution.

5.3.3 Filter Design by Frequency Sampling

Audio filter design for nonrecursive filter realizations by fast convolution can be carried out by the frequency sampling method. For linear phase systems we obtain

$$H(e^{j\Omega}) = A(e^{j\Omega})\, e^{-j(N_F-1/2)\Omega}, \tag{5.227}$$

where $A(e^{j\Omega})$ is a real-valued amplitude response and N_F is the length of the impulse response. The magnitude $|H(e^{j\Omega})|$ is calculated by sampling in the frequency domain at equidistant places

$$\frac{f}{f_S} = \frac{k}{N_F}, \quad k = 0, 1, \ldots, N_F - 1, \tag{5.228}$$

according to

$$|H(e^{j\Omega})| = A(e^{j2\pi k/N_F}), \quad k = 0, 1, \ldots, \frac{N_F}{2} - 1. \tag{5.229}$$

Hence, a filter can be designed by fulfilling conditions in the frequency domain. The linear phase is determined as

$$e^{-j\frac{N_F-1}{2}\Omega} = e^{-j2\pi(N_F-1/2)k/N_F} \tag{5.230}$$

$$= \cos\left(2\pi\frac{N_F-1}{2}\frac{k}{N_F}\right) - j\sin\left(2\pi\frac{N_F-1}{2}\frac{k}{N_F}\right), \tag{5.231}$$

$$k = 0, 1, \ldots, \frac{N_F}{2} - 1.$$

Owing to the real transfer function $H(z)$ for an even filter length, we have to fulfill

$$H\left(k = \frac{N_F}{2}\right) = 0 \wedge H(k) = H^*(N_F - k), \quad k = 0, 1, \ldots, \frac{N_F}{2} - 1. \tag{5.232}$$

This has to be taken into consideration while designing filters of even length N_F. The impulse response $h(n)$ is obtained through an N_F-point IDFT of the spectral sequence $H(k)$. This impulse response is extended with *zero-padding* to the length N and then transformed by an N-point DFT resulting in the spectral sequence $H(k)$ of the filter.

Example: For $N_F = 8$, $|H(k)| = 1$ ($k = 0, 1, 2, \ldots, 7$) and $|H(4)| = 0$, the group delay is $t_G = 3.5$. Figure 5.60 shows the amplitude, real part and imaginary part of the transfer function and the impulse response $h(n)$.

5.4 Multi-complementary Filter Bank

The subband processing of audio signals is mainly used in source coding applications for efficient transmission and storing. The basis for the subband decomposition is critically sampled filter banks [Vai93, Fli00]. These filter banks allow a perfect reconstruction of the input provided there is no processing within the subbands. They consist of an analysis filter bank for decomposing the signal in critically sampled subbands and a synthesis filter bank for reconstructing the broad-band output. The aliasing in the subbands is eliminated by the synthesis filter bank. Nonlinear methods are used for coding the subband signals. The reconstruction error of the filter bank is negligible compared with the errors due to the coding/decoding process. Using a critically sampled filter bank as a multi-band equalizer, multi-band dynamic range control or multi-band room simulation, the processing in the subbands leads to aliasing at the output. In order to avoid aliasing, a multi-complementary filter bank [Fli92, Zöl92, Fli00] is presented which enables an aliasing-free processing in the subbands and leads to a perfect reconstruction of the output. It allows a decomposition into octave frequency bands which are matched to the human ear.

5.4.1 Principles

Figure 5.61 shows an octave-band filter bank with *critical sampling*. It performs a successive low-pass/high-pass decomposition into half-bands followed by downsampling by a factor 2. The decomposition leads to the subbands Y_1 to Y_N (see Fig. 5.62). The transition

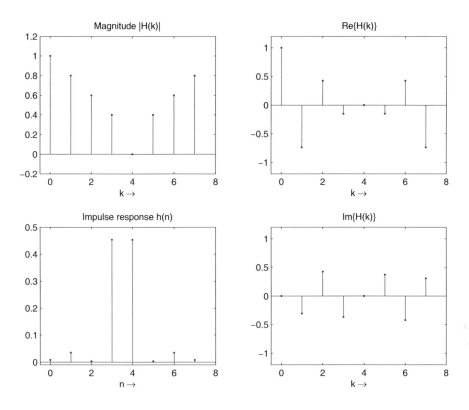

Figure 5.60 Filter design by frequency sampling (N_F even).

frequencies of this decomposition are given by

$$\Omega_{Ck} = \frac{\pi}{2} 2^{-k+1}, \quad k = 1, 2, \ldots, N - 1. \tag{5.233}$$

In order to avoid aliasing in subbands, a modified octave-band filter bank is considered which is shown in Fig. 5.63 for a two-band decomposition. The cutoff frequency of the modified filter bank is moved from $\frac{\pi}{2}$ to a lower frequency. This means that in downsampling the low-pass branch, no aliasing occurs in the transition band (e.g. cutoff frequency $\frac{\pi}{3}$). The broader high-pass branch cannot be downsampled. A continuation of the two-band decomposition described leads to the modified octave-band filter bank shown in Fig. 5.64. The frequency bands are depicted in Fig. 5.65 showing that besides the cutoff frequencies

$$\Omega_{Ck} = \frac{\pi}{3} 2^{-k+1}, \quad k = 1, 2, \ldots, N - 1, \tag{5.234}$$

the bandwidth of the subbands is reduced by a factor 2. The high-pass subband Y_1 is an exception.

The special low-pass/high-pass decomposition is carried out by a two-band complementary filter bank as shown in Fig. 5.66. The frequency responses of a decimation filter $H_D(z)$, interpolation filter $H_I(z)$ and kernel filter $H_K(z)$ are shown in Fig. 5.67.

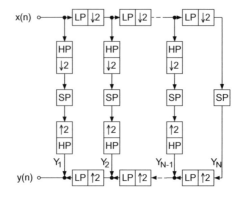

Figure 5.61 Octave-band QMF filter bank (SP = signal processing, LP = low-pass, HP = high-pass).

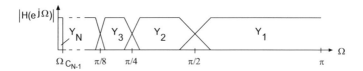

Figure 5.62 Octave-frequency bands.

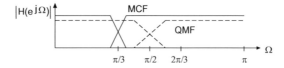

Figure 5.63 Two-band decomposition.

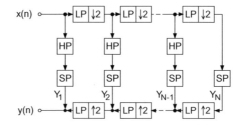

Figure 5.64 Modified octave-band filter bank.

The low-pass filtering of a signal $x_1(n)$ is done with the help of a decimation filter $H_D(z)$, the downsampler of factor 2 and the kernel filter $H_K(z)$ and leads to $y_2(2n)$. The Z-transform of $y_2(2n)$ is given by

$$Y_2(z) = \tfrac{1}{2}[H_D(z^{\frac{1}{2}})X_1(z^{\frac{1}{2}})H_K(z) + H_D(-z^{\frac{1}{2}})X_1(-z^{\frac{1}{2}})H_K(z)]. \qquad (5.235)$$

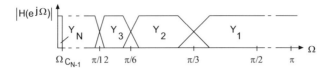

Figure 5.65 Modified octave decomposition.

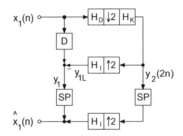

Figure 5.66 Two-band complementary filter bank.

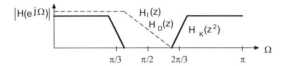

Figure 5.67 Design of $H_D(z)$, $H_I(z)$ and $H_K(z)$.

The interpolated low-pass signal $y_{1L}(n)$ is generated by upsampling by a factor 2 and filtering with the interpolation filter $H_I(z)$. The Z-transform of $y_{1L}(n)$ is given by

$$Y_{1L}(z) = Y_2(z^2)H_I(z) \tag{5.236}$$
$$= \underbrace{\tfrac{1}{2}H_D(z)H_I(z)H_K(z^2)}_{G_1(z)}\,X_1(z) + \underbrace{\tfrac{1}{2}H_D(-z)H_I(z)H_K(z^2)}_{G_2(z)}\,X_1(-z). $$

$$\tag{5.237}$$

The high-pass signal $y_1(n)$ is obtained by subtracting the interpolated low-pass signal $y_{1L}(n)$ from the delayed input signal $x_1(n - D)$. The Z-transform of the high-pass signal is given by

$$Y_1(z) = z^{-D}X_1(z) - Y_{1L}(z) \tag{5.238}$$
$$= [z^{-D} - G_1(z)]X_1(z) - G_2(z)X_1(-z). \tag{5.239}$$

The low-pass and high-pass signals are processed individually. The output signal $\hat{x}_1(n)$ is formed by adding the high-pass signal to the upsampled and filtered low-pass signal. With (5.237) and (5.239) the Z-transform of $\hat{x}_1(n)$ can be written as

$$\hat{X}_1(z) = Y_{1L}(z) + Y_1(z) = z^{-D}X_1(z). \tag{5.240}$$

Equation (5.240) shows the perfect reconstruction of the input signal which is delayed by D sampling units.

The extension to N subbands and performing the kernel filter using complementary techniques [Ram88, Ram90] leads to the multi-complementary filter bank as shown in Fig. 5.68. Delays are integrated in the high-pass (Y_1) and band-pass subbands (Y_2 to Y_{N-2}) in order to compensate the group delay. The filter structure consists of N horizontal stages. The kernel filter is implemented as a complementary filter in S vertical stages. The design of the latter will be discussed later. The vertical delays in the extended kernel filters (EKF$_1$ to EKF$_{N-1}$) compensate group delays caused by forming the complementary component. At the end of each of these vertical stages is the kernel filter H_K. With

$$z_k = z^{2^{-(k-1)}} \quad \text{and} \quad k = 1, \ldots, N, \tag{5.241}$$

the signals $\hat{X}_k(z_k)$ can be written as a function of the signals $X_k(z_k)$ as

$$\hat{\mathbf{X}} = \text{diag} \, [z_1^{-D_1} \quad z_2^{-D_2} \quad \ldots \quad z_N^{-D_N}] \mathbf{X}, \tag{5.242}$$

with

$$\hat{\mathbf{X}} = [\hat{X}_1(z_1) \quad \hat{X}_2(z_2) \quad \ldots \quad \hat{X}_N(z_N)]^T,$$
$$\mathbf{X} = [X_1(z_1) \quad X_2(z_2) \quad \ldots \quad X_N(z_N)]^T,$$

and with $k = N - l$ the delays are given by

$$D_{k=N} = 0, \tag{5.243}$$
$$D_{k=N-l} = 2D_{N-l+1} + D, \quad l = 1, \ldots, N - 1. \tag{5.244}$$

Perfect reconstruction of the input signal can be achieved if the horizontal delays D_{Hk} are given by

$$D_{H_{k=N}} = 0,$$
$$D_{H_{k=N-1}} = 0,$$
$$D_{H_{k=N-l}} = 2D_{N-l+1}, \quad l = 2, \ldots, N - 1.$$

The implementation of the extended vertical kernel filters is done by calculating complementary components as shown in Fig. 5.69. After upsampling, interpolating with a high-pass HP (Fig. 5.69b) and forming the complementary component, the kernel filter H_K with frequency response as in Fig. 5.69a becomes low-pass with frequency response as illustrated in Fig. 5.69c. The slope of the filter characteristic remains constant whereas the cutoff frequency is doubled. A subsequent upsampling with an interpolation high-pass (Fig. 5.69d) and complement filtering leads to the frequency response in Fig. 5.69e. With the help of this technique, the kernel filter is implemented at a reduced sampling rate. The cutoff frequency is moved to a desired cutoff frequency by using decimation/interpolation stages with complement filtering.

Computational Complexity. For an N-band multi-complementary filter bank with $N - 1$ decomposition filters where each is implemented by a kernel filter with S stages, the horizontal complexity is given by

$$HC = HC_1 + HC_2 \left(\frac{1}{2} + \frac{1}{4} + \cdots + \frac{1}{2^N} \right). \tag{5.245}$$

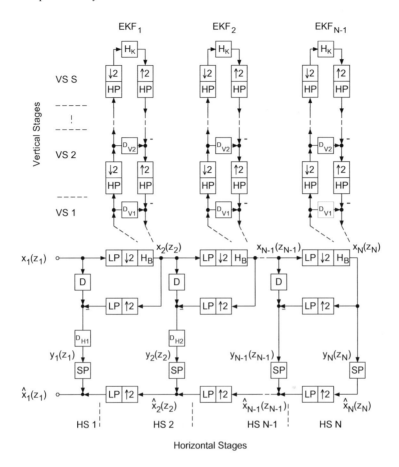

Figure 5.68 Multi-complementary filter bank.

HC_1 denotes the number of operations that are carried out at the input sampling rate. These operations occur in the horizontal stage HS_1 (see Fig. 5.68). HC_2 denotes the number of operations (horizontal stage HS_2) that are performed at half of the sampling rate. The number of operations in the stages from HS_2 to HS_N are approximately identical but are calculated at sampling rates that are successively halved.

The complexities VC_1 to VC_{N-1} of the vertical kernel filters EKF_1 to EKF_{N-1} are calculated as

$$VC_1 = \frac{1}{2}V_1 + V_2\left(\frac{1}{4} + \frac{1}{8} + \cdots + \frac{1}{2^{S+1}}\right),$$

$$VC_2 = \frac{1}{4}V_1 + V_2\left(\frac{1}{8} + \frac{1}{16} + \cdots + \frac{1}{2^{S+2}}\right) = \frac{1}{2}VC_1,$$

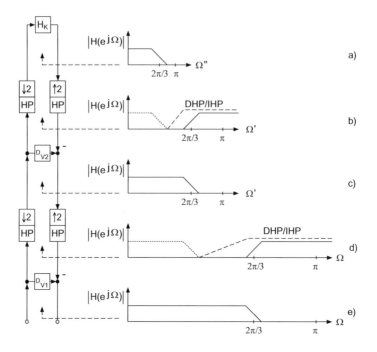

Figure 5.69 Multirate complementary filter.

$$VC_3 = \frac{1}{8}V_1 + V_2\left(\frac{1}{16} + \frac{1}{32} + \cdots + \frac{1}{2^{S+3}}\right) = \frac{1}{4}VC_1,$$

$$VC_{N-1} = \frac{1}{2^{N-1}}V_1 + V_2\left(\frac{1}{2^N} + \cdots + \frac{1}{2^{S+N-1}}\right) = \frac{1}{2^{N-1}}VC_1,$$

where V_1 depicts the complexity of the first stage VS_1 and V_2 is the complexity of the second stage VS_2 (see Fig. 5.68). It can be seen that the total vertical complexity is given by

$$VC = VC_1\left(1 + \frac{1}{2} + \frac{1}{4} + \cdots + \frac{1}{2^{N-1}}\right). \tag{5.246}$$

The upper bound of the total complexity results is the sum of horizontal and vertical complexities and can be written as

$$C_{\text{tot}} = HC_1 + HC_2 + 2VC_1. \tag{5.247}$$

The total complexity C_{tot} is independent of the number of frequency bands N and vertical stages S. This means that for real-time implementation with finite computation power, any desired number of subbands with arbitrarily narrow transition bands can be implemented!

5.4.2   Example: Eight-band Multi-complementary Filter Bank

In order to implement the frequency decomposition into the eight bands shown in Fig. 5.70, the multirate filter structure of Fig. 5.71 is employed. The individual parts of the system provide means of downsampling (D = decimation), upsampling (I = interpolation), kernel filtering (K), signal processing (SP), delays (N_1 = Delay 1, N_2 = Delay 2) and group delay compensation M_i in the ith band. The frequency decomposition is carried out successively from the highest to the lowest frequency band. In the two lowest frequency bands, a compensation for group delay is not required. The slope of the filter response can be adjusted with the kernel complementary filter structure shown in Fig. 5.72 which consists of one stage. The specifications of an eight-band equalizer are listed in Table 5.11. The stop-band attenuation of the subband filters is chosen to be 100 dB.

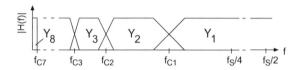

Figure 5.70 Modified octave decomposition of the frequency band.

Table 5.11 Transition frequencies f_{Ci} and transition bandwidths TB in an eight-band equalizer.

f_S [kHz]	f_{C1} [Hz]	f_{C2} [Hz]	f_{C3} [Hz]	f_{C4} [Hz]	f_{C5} [Hz]	f_{C6} [Hz]	f_{C7} [Hz]
44.1	7350	3675	1837.5	918.75	459.375	≈230	≈115
TB [Hz]	1280	640	320	160	80	40	20

Filter Design

The design of different decimation and interpolation filters is mainly determined by the transition bandwidth and the stop-band attenuation for the lower frequency band. As an example, a design is made for an eight-band equalizer. The kernel complementary filter structure for both lower frequency bands is illustrated in Fig. 5.72. The design specifications for the kernel low-pass, decimation and interpolation filters are presented in Fig. 5.73.

Kernel Filter Design. The transition bandwidth of the kernel filter is known if the transition bandwidth is given for the lower frequency band. This kernel filter must be designed for a sampling rate of $f_S'' = 44100/(2^8)$. For a given transition bandwidth f_{TB} at a frequency $f'' = f_S''/3$, the normalized pass-band frequency is

$$\frac{\Omega_{Pb}''}{2\pi} = \frac{f'' - f_{TB}/2}{f_S''} \tag{5.248}$$

and the normalized stop-band frequency

$$\frac{\Omega_{Sb}''}{2\pi} = \frac{f'' + f_{TB}/2}{f_S''}. \tag{5.249}$$

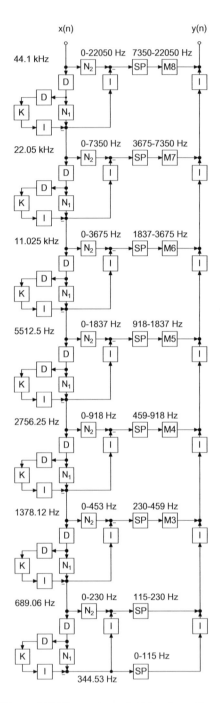

Figure 5.71 Linear phase eight-band equalizer.

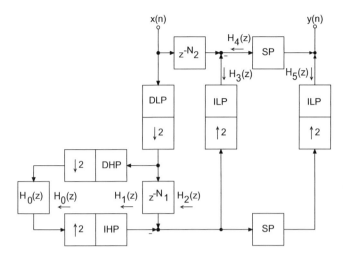

Figure 5.72 Kernel complementary filter structure.

With the help of these parameters the filter can be designed. Making use of the Parks–McClellan program, the frequency response shown in Fig. 5.74 is obtained for a transition bandwidth of $f_{TB} = 20$ Hz. The necessary filter length for a stop-band attenuation of 100 dB is 53 taps.

Decimation and Interpolation High-pass Filter. These filters are designed for a sampling rate of $f'_S = 44100/(2^7)$ and are half-band filters as illustrated in Fig. 5.73. First a low-pass filter is designed, followed by a high-pass to low-pass transformation. For a given transition bandwidth f_{TB}, the normalized pass-band frequency is

$$\frac{\Omega'_{Pb}}{2\pi} = \frac{f'' + f_{TB}/2}{f'_S} \tag{5.250}$$

and the normalized stop-band frequency is given by

$$\frac{\Omega'_{Sb}}{2\pi} = \frac{2f'' - f_{TB}/2}{f'_S}. \tag{5.251}$$

With these parameters the design of a half-band filter is carried out. Figure 5.75 shows the frequency response. The necessary filter length for a stop-band attenuation of 100 dB is 55 taps.

Decimation and Interpolation Low-pass Filter. These filters are designed for a sampling rate of $f_S = 44100/(2^6)$ and are also half-band filters. For a given transition bandwidth f_{TB}, the normalized pass-band frequency is

$$\frac{\Omega_{Pb}}{2\pi} = \frac{2f'' + f_{TB}/2}{f_S} \tag{5.252}$$

and the normalized stop-band frequency is given by

$$\frac{\Omega_{Sb}}{2\pi} = \frac{4f'' - f_{TB}/2}{f_S}. \tag{5.253}$$

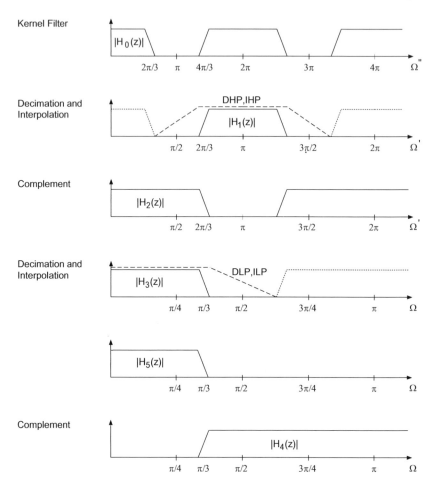

Figure 5.73 Decimation and interpolation filters.

With these parameters the design of a half-band filter is carried out. Figure 5.76 shows the frequency response. The necessary filter length for a stop-band attenuation of 100 dB is 43 taps. These filter designs are used in every decomposition stage so that the transition frequencies and bandwidths are obtained as listed in Table 5.11.

Memory Requirements and Latency Time. The memory requirements depend directly on the transition bandwidth and the stop-band attenuation. Here, the memory operations for the actual kernel, decimation and interpolation filters have to be differentiated from the group delay compensations in the frequency bands. The compensating group delay N_1 for decimation and interpolation high-pass filters of order $O_{\text{DHP/IHP}}$ is calculated with the help of the kernel filter order O_{KF} according to

$$N_1 = O_{\text{KF}} + O_{\text{DHP/IHP}}. \tag{5.254}$$

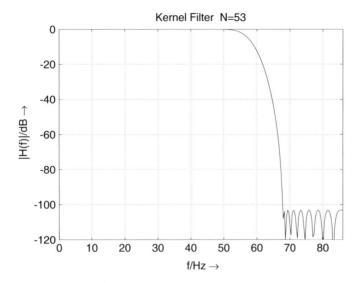

Figure 5.74 Kernel low-pass filter with a transition bandwidth of 20 Hz.

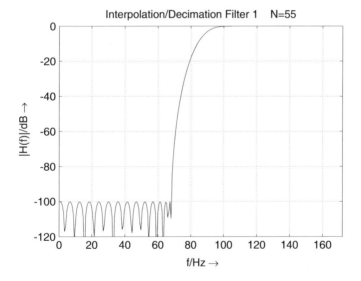

Figure 5.75 Decimation and interpolation high-pass filter.

The group delay compensation N_2 for the decimation and interpolation low-pass filters of order $O_{\mathrm{DLP/ILP}}$ is given by

$$N_2 = 2N_1 + O_{\mathrm{DLP/ILP}}. \tag{5.255}$$

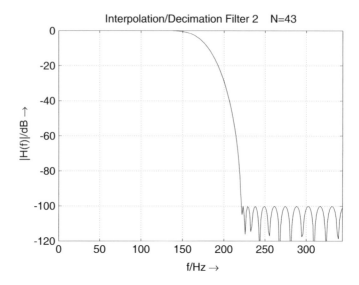

Figure 5.76 Decimation and interpolation low-pass filter.

The delays $M_3, \ldots, M_8$ in the individual frequency bands are calculated recursively starting from the two lowest frequency bands:

$$M_3 = 2N_2, \quad M_4 = 6N_2, \quad M_5 = 14N_2,$$
$$M_6 = 30N_2, \quad M_7 = 62N_2, \quad M_8 = 126N_2.$$

The memory requirements per decomposition stage are listed in Table 5.12. The memory for the delays can be computed by $\sum_i M_i = 240N_2$. The latency time (delay) is given by $t_D = (M_8/44100)10^3$ ms ($t_D = 725$ ms).

Table 5.12 Memory requirements.

Kernel filter	O_{KF}
DHP/IHP	$2 \cdot O_{DHP/IHP}$
DLP/ILP	$3 \cdot O_{DLP/ILP}$
N_1	$O_{KF} + O_{DHP/IHP}$
N_2	$2 \cdot N_1 + O_{DLP/ILP}$

5.5 Java Applet – Audio Filters

The applet shown in Fig. 5.77 demonstrates audio filters. It is designed for a first insight into the perceptual effect of filtering an audio signal. Besides the different filter types and their acoustical effect, the applet offers a first insight into the logarithmic behavior of loudness and frequency resolution of our human acoustical perception.

The following filter functions can be selected on the lower right of the graphical user interface:

- Low-/high-pass filter (LP/HP) with control parameter
    - cutoff frequency f_c in hertz (lower horizontal slider)
    - all frequencies above (LP) or below (HP) the cutoff frequency are attenuated according to the shown frequency response.

- Low/high-frequency shelving filter (LFS/HFS) with control parameters
    - cutoff frequency f_c in hertz (lower horizontal slider)
    - boost/cut in dB (left vertical slider with + for boost or − for cut)
    - all frequencies below (LFS) or above (HFS) the cutoff frequency are boosted/cut according to the selected boost/cut.

- Peak filter with control parameters
    - center frequency f_c in hertz (lower horizontal slider)
    - boost/cut in dB (left vertical slider with + for boost or − for cut)
    - Q-factor $Q = f_c/f_b$ (right vertical slider), which controls the bandwidth f_b of the boost/cut around the adjusted center frequency f_c. Lower Q-factor means wider bandwidth.
    - the peak filter boosts/cuts the center frequency with a bandwidth adjusted by the Q-factor.

The center window shows the frequency response (filter gain versus frequency) of the selected filter functions. You can choose between a linear and a logarithmic frequency axis.

You can choose between two predefined audio files from our web server (*audio1.wav* or *audio2.wav*) or your own local wav file to be processed [Gui05].

5.6   Exercises

1. Design of Recursive Audio Filters

1. How can we design a low-frequency shelving filter? Which parameters define the filter? Explain the control parameters.

2. How can we derive a high-frequency shelving filter? Which parameters define the filter?

3. What is the difference between first- and second-order shelving filters.

4. How can we design a peak filter? Which parameters define the filter? What is the filter order? Explain the control parameters. Explain the Q-factor.

5. How do we derive the digital transfer function?

6. Derive the digital transfer functions for the first-order shelving filters.

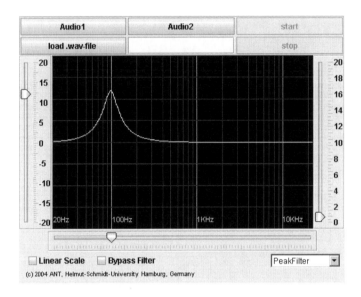

Figure 5.77 Java applet – audio filters.

2. Parametric Audio Filters

1. What is the basic idea of parametric filters?

2. What is the difference between the Regalia and the Zölzer filter structures? Count the number of multiplications and additions for both filter structures.

3. Derive a signal flow graph for first- and second-order parametric Zölzer filters with a direct-form implementation of the all-pass filters.

4. Is there a complete decoupling of all control parameters for boost and cut? Which parameters are decoupled?

3. Shelving Filter: Direct Form

Derive a first-order low shelving filter from a purely band-limiting first-order low-pass filter. Use a bilinear transform and give the transfer function of the low shelving filter.

1. Write down what you know about the filter coefficients and calculate the poles/zeros as functions of V_0 and T. What gain factor do you have if $z = \pm 1$?

2. What is the difference between purely band-limiting filters and the shelving filter?

3. How can you describe the boost and cut effect related to poles/zeros of the filter?

4. How do we get a transfer function for the boost case from the cut case?

5. How do we go from a low shelving filter to a high shelving filter?

4. Shelving Filter: All-pass Form

Implement a first-order high shelving filter for the boost and cut cases with the sampling rate $f_S = 44.1$ kHz, cutoff frequency $f_c = 10$ kHz and gain $G = 12$ dB.

1. Define the all-pass parameters and coefficients for the boost and cut cases.

2. Derive from the all-pass decomposition the complete transfer function of the shelving filter.

3. Using Matlab, give the magnitude frequency response for boost and cut. Show the result for the case where a boost and cut filter are in a series connection.

4. If the input signal to the system is a unit impulse, give the spectrum of the input and out signal for the boost and cut cases. What result do you expect in this case when boost and cut are again cascaded?

5. Quantization of Filter Coefficients

For the quantization of the filter coefficients different methods have been proposed: direct form, Gold and Rader, Kingsbury and Zölzer.

1. What is the motivation behind this?

2. Plot a pole distribution using the quantized polar representation of a second-order IIR filter

$$H(z) = \frac{N(z)}{1 - 2r \cos \varphi z^{-1} + r^2 z^{-2}}.$$

6. Signal Quantization inside the Audio Filter

Now we combine coefficient and signal quantization.

1. Design a digital high-pass filter (second-order IIR), with a cutoff frequency $f_c = 50$ Hz. (Use the Butterworth, Chebyshev or elliptic design methods implemented in Matlab.)

2. Quantize the signal only when it leaves the accumulator (i.e. before it is saved in any state variable).

3. Now quantize the coefficients (direct form), too.

4. Extend your quantization to every arithmetic operation (i.e. after each addition/ multiplication).

7. Quantization Effects in Recursive Audio Filters

1. Why is the quantization of signals inside a recursive filter of special interest?

2. Derive the noise transfer function of the second-order direct-form filter. Apply a first- and second-order noise shaping to the quantizer inside the direct-form structure and discuss its influence. What is the difference between second-order noise shaping and double-precision arithmetic?

3. Write a Matlab implementation of a second-order filter structure for quantization and noise shaping.

8. Fast Convolution

For an input sequence $x(n)$ of length $N_1 = 500$ and the impulse response $h(n)$ of length $N_2 = 31$, perform the discrete-time convolution.

1. Give the discrete-time convolution sum formula.

2. Using Matlab, define $x(n)$ as a sum of two sinusoids and derive $h(n)$ with Matlab function `remez(..)`.

3. Realize the filter operation with Matlab using:

    • the function `conv(x,h)`
    • the sample-by-sample convolution sum method
    • the FFT method
    • the FFT with overlap-add method.

4. Describe FIR filtering with the fast convolution technique. What conditions do the input signal and the impulse responses have to fulfill if convolution is performed by equivalent frequency-domain processing?

5. What happens if input signal and impulse response are as long as the FFT transform length?

6. How can we perform the IFFT by the FFT algorithm?

7. Explain the processing steps

    • for a segmentation of the input signal into blocks and fast convolution;
    • for a stereo signal by the fast convolution technique;
    • for the segmentation of the impulse response.

8. What is the processing delay of the fast convolution technique?

9. Write a Matlab program for fast convolution.

10. How does quantization of the signal influence the roundoff noise behavior of an FIR filter?

9. FIR Filter Design by Frequency Sampling

1. Why is frequency sampling an important design method for audio equalizers? How do we sample magnitude and phase response?

2. What is the linear phase frequency response of a system? What is the effect on an input signal passing through such a system?

3. Explain the derivation of the magnitude and phase response for a linear phase FIR filter.

4. What is the condition for a real-valued impulse response of even length N? What is the group delay?

5. Write a Matlab program for the design of an FIR filter and verify the example in the book.

6. If the desired frequency response is an ideal low-pass filter of length $N_F = 31$ with cutoff frequency $\Omega_c = \pi/2$, derive the impulse response of this system. What will the result be for $N_F = 32$ and $\Omega_c = \pi$?

10. Multi-complementary Filter Bank

1. What is an octave-spaced frequency splitting and how can we design a filter bank for that task?

2. How can we perform aliasing-free subband processing? How can we achieve narrow transition bands for a filter bank? What is the computational complexity of an octave-spaced filter bank?

References

[Abu79] A. I. Abu-El-Haija, A. M. Peterson: *An Approach to Eliminate Roundoff Errors in Digital Filters*, IEEE Trans. ASSP, Vol. 3, pp. 195–198, April 1979.

[Ave71] E. Avenhaus: *Zum Entwurf digitaler Filter mit minimaler Speicherwortlänge für Koeffizienten und Zustandsgrössen*, Ausgewählte Arbeiten über Nachrichtensysteme, No. 13, Prof. Dr.-Ing. W. Schüssler, Ed., Erlangen 1971.

[Bar82] C. W. Barnes: *Error Feedback in Normal Realizations of Recursive Digital Filters*, IEEE Trans. Circuits and Systems, pp. 72–75, January 1982.

[Bom85] B. W. Bomar: *New Second-Order State-Space Structures for Realizing Low Roundoff Noise Digital Filters*, IEEE Trans. ASSP, Vol. 33, pp. 106–110, Feb. 1985.

[Bri94] R. Bristow-Johnson: *The Equivalence of Various Methods of Computing Biquad Coefficients for Audio Parametric Equalizers*, Proc. 97th AES Convention, Preprint No. 3906, San Francisco, November 1994.

[Cad87]   J. A. Cadzow: *Foundations of Digital Signal Processing and Data Analysis*, New York: Macmillan, 1987.

[Cha78]   T. L. Chang: *A Low Roundoff Noise Digital Filter Structure*, Proc. Int. Symp. on Circuits and Systems, pp. 1004–1008, May 1978.

[Cla00]   R. J. Clark, E. C. Ifeachor, G. M. Rogers, P. W. J. Van Eetvelt: *Techniques for Generating Digital Equalizer Coefficients*, J. Audio Eng. Soc., Vol. 48, pp. 281–298, April 2000.

[Cre03]   L. Cremer, M. Möser: *Technische Akustik*, Springer-Verlag, Berlin, 2003.

[Dat97]   J. Dattorro: *Effect Design – Part 1: Reverberator and Other Filters*, J. Audio Eng. Soc., Vol. 45, pp. 660–684, September 1997.

[Din95]   Yinong Ding, D. Rossum: *Filter Morphing of Parametric Equalizers and Shelving Filters for Audio Signal Processing*, J. Audio Eng. Soc., Vol. 43, pp. 821–826, October 1995.

[Duh88]   P. Duhamel, B. Piron, J. Etcheto: *On Computing the Inverse DFT*, IEEE Trans. Acoust., Speech, Signal Processing, Vol. 36, pp. 285–286, February 1988.

[Ege96]   G. P. M. Egelmeers, P. C. W. Sommen: *A New Method for Efficient Convolution in Frequency Domain by Nonuniform Partitioning for Adaptive Filtering*, IEEE Trans. Signal Processing, Vol. 44, pp. 3123–3192, December 1996.

[Ell82]   D. F. Elliott, K. R. Rao: *Fast Transforms: Algorithms, Analyses, Applications*, Academic Press, New York, 1982.

[Fet72]   A. Fettweis: *On the Connection Between Multiplier Wordlength Limitation and Roundoff Noise in Digital Filters*, IEEE Trans. Circuit Theory, pp. 486–491, September 1972.

[Fli92]   N. J. Fliege, U. Zölzer: *Multi-complementary Filter Bank: A New Concept with Aliasing-Free Subband Signal Processing and Perfect Reconstruction*, Proc. EUSIPCO-92, pp. 207–210, Brussels, August 1992.

[Fli93a]   N.J. Fliege, U. Zölzer: *Multi-complementary Filter Bank*, Proc. ICASSP-93, pp. 193–196, Minneapolis, April 1993.

[Fli00]   N. Fliege: *Multirate Digital Signal Processing*, John Wiley & Sons, Ltd, Chichester, 2000.

[Fon01]   F. Fontana, M. Karjalainen: *Magnitude-Complementary Filters for Dynamic Equalization*, Proc. of the DAFX-01, Limerick, pp. 97–101, December 2001.

[Fon03]   F. Fontana, M. Karjalainen: *A Digital Bandpass/Bandstop Complementary Equalization Filter with Independent Tuning Characteristics*, IEEE Signal Processing Letters, Vol. 10, No. 4, pp. 119–122, April 2003.

[Gar95]   W. G. Gardner: *Efficient Convolution Without Input-Output Delay*, J. Audio Eng. Soc., Vol. 43, pp. 127–136, 1995.

[Garc02] G. Garcia: *Optimal Filter Partition for Efficient Convolution with Short Input/Output Delay*, Proc. 113th AES Convention, Preprint No. 5660, Los Angeles, October 2002.

[Gol67]   B. Gold, C. M. Rader: *Effects of Parameter Quantization on the Poles of a Digital Filter*, Proc. IEEE, Vol. 55, pp. 688–689, May 1967.

[Gui05]   M. Guillemard, C. Ruwwe, U. Zölzer: *J-DAFx – Digital Audio Effects in Java*, Proc. 8th Int. Conference on Digital Audio Effects (DAFx-05), pp. 161–166, Madrid, 2005.

[Har93]   F. J. Harris, E. Brooking: *A Versatile Parametric Filter Using an Imbedded All-Pass Sub-filter to Independently Adjust Bandwidth, Center Frequency and Boost or Cut*, Proc. 95th AES Convention, Preprint No. 3757, San Francisco, 1993.

[Här98]   A. Härmä: *Implementation of Recursive Filters Having Delay Free Loops*, Proc. IEEE Int. Conf. Acoustics, Speech, and Signal Processing (ICASSP'98), Vol. 3, pp. 1261–1264, Seattle, 1998.

[Hol06a] M. Holters, U. Zölzer: *Parametric Recursive Higher-Order Shelving Filters*, Jahrestagung Akustik DAGA, Braunschweig, March 2006.

[Hol06b] M. Holters, U. Zölzer: *Parametric Recursive Higher-Order Shelving Filters*, Proc. 120th Conv. Audio Eng. Soc., Preprint No. 6722, Paris, 2006.

[Hol06c] M. Holters, U. Zölzer: *Parametric Higher-Order Shelving Filters*, Proc. EUSIPCO-06, Florence, 2006.

[Hol06d] M. Holters, U. Zölzer: *Graphic Equalizer Design Using Higher-Order Recursive Filters*, Proc. of the 9th Int. Conference on Digital Audio Effects (DAFx-06), pp. 37–40, Montreal, September 2006.

[Jur64]   E. I. Jury: *Theory and Application of the z-Transform Method*, John Wiley & Sons, Inc., New York, 1964.

[Kei04]   F. Keiler, U. Zölzer, *Parametric Second- and Fourth-Order Shelving Filters for Audio Applications*, Proc. of IEEE 6th International Workshop on Multimedia Signal Processing, Siena, September–October 2004.

[Kin72]   N. G. Kingsbury: *Second-Order Recursive Digital Filter Element for Poles Near the Unit Circle and the Real z-Axis*, Electronic Letters, Vol. 8, pp. 155–156, March 1972.

[Moo83]   J. A. Moorer: *The Manifold Joys of Conformal Mapping*, J. Audio Eng. Soc., Vol. 31, pp. 826–841, 1983.

[Mou90]   J. N. Mourjopoulos, E. D. Kyriakis-Bitzaros, C. E. Goutis: *Theory and Real-Time Implementation of Time-Varying Digital Audio Filters*, J. Audio Eng. Soc., Vol. 38, pp. 523–536, July/August 1990.

[Mul76]   C. T. Mullis, R. A. Roberts: *Synthesis of Minimum Roundoff Noise Fixed Point Digital Filters*, IEEE Trans. Circuits and Systems, pp. 551–562, September 1976.

[Mül99]   C. Müller-Tomfelde: *Low Latency Convolution for Real-time Application*, In Proc. of the AES 16th International Conference: Spatial Sound Reproduction, pp. 454–460, Rovaniemi, Finland, April 1999.

[Mül01]   C. Müller-Tomfelde: *Time-Varying Filter in Non-uniform Block Convolution*, Proc. of the COST G-6 Conference on Digital Audio Effects (DAFX-01), Limerick, December 2001.

[Orf96a] S. J. Orfanidis: *Introduction to Signal Processing*, Prentice Hall, Englewood Cliffs, NJ, 1996.

[Orf96b] S. J. Orfanidis: *Digital Parametric Equalizer Design with Prescribed Nyquist-Frequency Gain*, Proc. 101st AES Convention, Preprint No. 4361, Los Angeles, November 1996.

[Orf05]   S. J. Orfanidis: *High-Order Digital Parametric Equalizer Design*, J. Audio Eng. Soc., Vol. 53, pp. 1026–1046, 2005.

[Rab88]   R. Rabenstein: *Minimization of Transient Signals in Recursive Time-Varying Digital Filters*, Circuits, Systems, and Signal Processing, Vol. 7, No. 3, pp. 345–359, 1988.

[Ram88]   T. A. Ramstad, T. Saramäki: *Efficient Multirate Realization for Narrow Transition-Band FIR Filters*, Proc. IEEE Int. Symp. on Circuits and Syst., pp. 2019–2022, Espoo, Finland, June 1988.

[Ram90]   T. A. Ramstad, T. Saramäki: *Multistage, Multirate FIR Filter Structures for Narrow Transition-Band Filters*, Proc. IEEE Int. Symp. on Circuits and Syst., pp. 2017–2021, New Orleans, May 1990.

[Reg87]   P. A. Regalia, S. K. Mitra: *Tunable Digital Frequency Response Equalization Filters*, IEEE Transactions on Acoustics, Speech, and Signal Processing, Vol. 35, No. 1, pp. 118–120, January 1987.

[Sha92]   D. J. Shpak: *Analytical Design of Biquadratic Filter Sections for Parametric Filters*, J. Audio Eng. Soc., Vol. 40, pp. 876–885, November 1992.

[Soo90]   J. S. Soo, K. K. Pang: *Multidelay Block Frequency Domain Adaptive Filter*, IEEE Transactions on Acoustics, Speech, and Signal Processing, Vol. 38, No. 2, pp. 373–376, February 1990.

[Sor87]   H. V. Sorensen, D. J. Jones, M. T. Heideman, C. S. Burrus: *Real-Valued Fast Fourier Transform Algorithms*, IEEE Trans. ASSP, Vol. 35, No. 6, pp. 849–863, June 1987.

[Sri77]   A. B. Sripad, D. L. Snyder: *A Necessary and Sufficient Condition for Quantization Errors to be Uniform and White*, IEEE Trans. ASSP, Vol. 25, pp. 442–448, October 1977.

[Tra77]   Tran-Thong, B. Liu: *Error Spectrum Shaping in Narrow Band Recursive Filters*, IEEE Trans. ASSP, pp. 200–203, April 1977.

[Vai93]   P. P. Vaidyanathan: *Multirate Systems and Filter Banks*, Prentice Hall, Englewood Cliffs, NJ, 1993.

[Väl98]   V. Välimäki, T. I. Laakso: *Suppression of Transients in Time-Varying Recursive Filters for Audio Signals*, Proc. IEEE Int. Conf. Acoustics, Speech, and Signal Processing (ICASSP'98), Vol. 6, pp. 3569–3572, Seattle, 1998.

[Whi86]   S. A. White: *Design of a Digital Biquadratic Peaking or Notch Filter for Digital Audio Equalization*, J. Audio Eng. Soc., Vol. 34, pp. 479–483, 1986.

[Zöl89]   U. Zölzer: *Entwurf digitaler Filter für die Anwendung im Tonstudiobereich*, Wissenschaftliche Beiträge zur Nachrichtentechnik und Signalverarbeitung, TU Hamburg-Harburg, June 1989.

[Zöl90]   U. Zölzer: *A Low Roundoff Noise Digital Audio Filter*, Proc. EUSIPCO-90, pp. 529–532, Barcelona, 1990.

[Zöl92]   U. Zölzer, N. Fliege: *Logarithmic Spaced Analysis Filter Bank for Multiple Loudspeaker Channels*, Proc. 93rd AES Convention, Preprint No. 3453, San Francisco, 1992.

[Zöl93]   U. Zölzer, B. Redmer, J. Bucholtz: *Strategies for Switching Digital Audio Filters*, Proc. 95th AES Convention, Preprint No. 3714, New York, October 1993.

[Zöl94]   U. Zölzer: *Roundoff Error Analysis of Digital Filters*, J. Audio Eng. Soc., Vol. 42, pp. 232–244, April 1994.

[Zöl95]   U. Zölzer, T. Boltze: *Parametric Digital Filter Structures*, Proc. 99th AES Convention, Preprint No. 4099, New York, October 1995.

Chapter 6

Room Simulation

Room simulation artificially reproduces the acoustics of a room. The foundations of room acoustics are found in [Cre78, Kut91]. Room simulation is mainly used for post-processing signals in which a microphone is located in the vicinity of an instrument or a voice. The direct signal, without additional room impression, is mapped to a certain acoustical room, for example a concert hall or a church. In terms of signal processing, the post-processing of an audio signal with room simulation corresponds to the convolution of the audio signal with a room impulse response.

6.1 Basics

6.1.1 Room Acoustics

The room impulse response between two points in a room can be classified as shown in Fig. 6.1. The impulse response consists of the direct signal, early reflections (from walls) and subsequent reverberation. The number of early reflections continuously increases with time and leads to a random signal with exponential decay called subsequent reverberation. The *reverberation time* (decrease in sound pressure level by 60 dB) can be calculated, using the geometry of the room and the partial areas that absorb sound in the room, from

$$T_{60} = 0.163 \frac{V}{\alpha S} = \frac{0.163}{[\text{m/s}]} \frac{V}{\sum_n \alpha_n S_n} \tag{6.1}$$

where T_{60} is the reverberation time (in s), V the volume of the room (m^3), S_n the partial areas (m^2) and α_n the absorption coefficient of partial area S_n.

The geometry of the room also determines the eigenfrequencies of a three-dimensional rectangular room:

$$f_e = \frac{c}{2} \sqrt{\left(\frac{n_x}{l_x}\right)^2 + \left(\frac{n_y}{l_y}\right)^2 + \left(\frac{n_z}{l_z}\right)^2}, \tag{6.2}$$

where n_x, n_y, n_z are integer number of half waves (0, 1, 2, ...), l_x, l_y, l_z are dimensions of a rectangular room, and c is the velocity of sound.

Digital Audio Signal Processing   Second Edition   Udo Zölzer
© 2008 John Wiley & Sons, Ltd

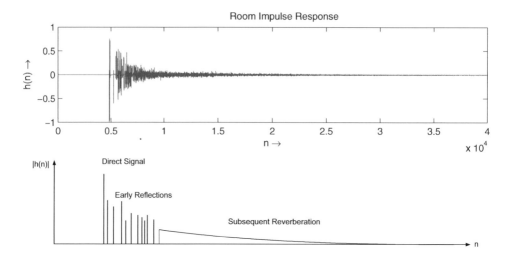

Figure 6.1 Room impulse response $h(n)$ and simplified decomposition into direct signal, early reflections and subsequent reverberation (with $|h(n)|$).

For larger rooms, the eigenfrequencies start from very low frequencies. In contrast, the lowest eigenfrequencies of smaller rooms are shifted toward higher frequencies. The mean frequency between two extrema of the frequency response of a large room is approximately inversely proportional to the reverberation time [Schr87]:

$$\Delta f \sim 1/T_{60}. \tag{6.3}$$

The distance between two eigenfrequencies decreases with increasing number of half waves. Above a *critical frequency*

$$f_c > 4000\sqrt{T_{60}/V}, \tag{6.4}$$

the density of eigenfrequencies becomes so large that they overlap each other [Schr87].

6.1.2   Model-based Room Impulse Responses

The methods for analytically determining a room impulse response are based on the ray tracing model [Schr70] or image model [All79]. In the case of the ray tracing model, a point source with radial emission is assumed. The path length of rays and the absorption coefficients of walls, roofs and floors are used to determine the room impulse response (see Fig. 6.2). For the image model, image rooms with secondary image sources are formed which in turn have further image rooms and image sources. The summation of all image sources with corresponding delays and attenuations provides the estimated room impulse response. Both methods are applied in room acoustics to get insight into the acoustical properties when planning concert halls, theaters, etc.

a) Ray Tracing   b) Virtual Image Sources

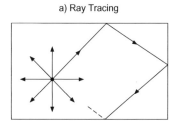

Figure 6.2 Model-based methods for calculating room impulse responses.

6.1.3 Measurement of Room Impulse Responses

The direct measurement of a room impulse response is carried out by impulse excitation. Better measurement results are obtained by correlation measurement of room impulse responses by using pseudo-random sequences as the excitation signal. Pseudo-random sequences can be generated by feedback shift registers [Mac76]. The pseudo-random sequence is periodic with period $L = 2^N - 1$, where N is the number of states of the shift register. The autocorrelation function (ACF) of such a random sequence is given by

$$r_{XX}(n) = \begin{cases} a^2, & n = 0, L, 2L, \ldots, \\ \dfrac{-a^2}{L}, & \text{elsewhere,} \end{cases} \qquad (6.5)$$

where a is the maximum value of the pseudo-random sequence. The ACF also has a period L. After going through a DA converter, the pseudo-random signal is fed through a loudspeaker into a room (see Fig. 6.3).

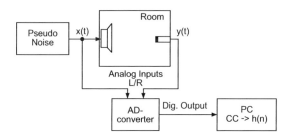

Figure 6.3 Measurement of room impulse response with pseudo-random signal $x(t)$.

At the same time, the pseudo-random signal and the room signal captured by a microphone are recorded on a personal computer. The impulse response is obtained with the cyclic cross-correlation

$$r_{XY}(n) = r_{XX}(n) * h(n) \approx \tilde{h}(n). \qquad (6.6)$$

For the measurement of room impulse responses it has to be borne in mind that the periodic length of the pseudo-random sequence must be longer than the length of the room impulse response. Otherwise, aliasing in the periodic cross-correlation $r_{XY}(n)$ (see Fig. 6.4) occurs.

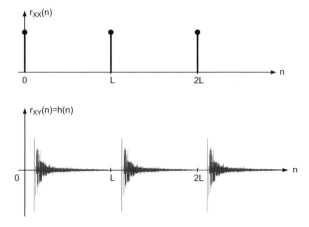

Figure 6.4 Periodic autocorrelation of pseudo-random sequence and periodic cross-correlation.

To improve the signal-to-noise ratio of the measurement, the average of several periods of the cross-correlation is calculated.

Sine sweep measurements [Far00, Mül01, Sta02] are based on a chirp signal $x_S(t)$ of length T_C and an inverse signal $x_{S_{inv}}(t)$, where both satisfy the condition

$$x_S(t) * x_{S_{inv}}(t) = \delta(t - T_C). \tag{6.7}$$

The chirp signal can be applied to the room by a loudspeaker and then a signal inside the room $y(t) = x_S(t) * h(n)$ is recorded. By performing the convolution of the received signal $y(t)$ with the inverse signal $x_{S_{inv}}(t)$ one obtains the impulse response from

$$y(t) * x_{S_{inv}}(t) = x_S(t) * h(n) * x_{S_{inv}}(t) = h(t - T_C). \tag{6.8}$$

6.1.4 Simulation of Room Impulse Responses

The methods just described provide means for calculating the impulse response from the geometry of a room and for measuring the impulse response of a real room. The reproduction of such an impulse response is basically possible with the help of the *fast convolution* method as described in Chapter 5. The ear signals at a listening position inside the room are computed by

$$y_L(n) = \sum_{k=0}^{N-1} x(k) \cdot h_L(n - k), \tag{6.9}$$

$$y_R(n) = \sum_{k=0}^{N-1} x(k) \cdot h_R(n - k), \tag{6.10}$$

where $h_L(n)$ and $h_R(n)$ are the measured impulse responses between the source inside the room, which generates the signal $x(n)$, and a dummy head with two ear microphones. Special implementations of fast convolution with low latency are described in [Soo90,

Gar95, Rei95, Ege96, Mül99, Joh00, Mül01, Garc02] and a hybrid approach based on convolution and recursive filters can be found in [Bro01]. Investigations regarding fast convolution with sparse psychoacoustic-based room impulse responses are discussed in [Iid95, Lee03a, Lee03b].

In the following sections we will consider special approaches for early reflections and subsequent reverberation, which allow a parametric adjustment of all relevant parameters of a room impulse response. With this approach an accurate room impulse response is not possible, but with a moderate computational complexity a satisfying solution from an acoustic point of view can be achieved. In Section 6.4 an efficient implementation of the convolutions (6.9) and (6.10) with a multirate signal processing approach [Zöl90, Sch92, Sch93, Sch94] is discussed.

6.2 Early Reflections

Early reflections decisively affect room perception. *Spatial impression* is produced by early reflections which reach the listener laterally. The significance of lateral reflections in creating *spatial impression* was investigated by Barron [Bar71, Bar81]. Fundamental investigations of concert halls and their different acoustics are described by Ando [And90].

6.2.1 Ando's Investigations

The results of the investigations by Ando are summarized in the following:

- Preferred *delay time of a single reflection*: with the ACF of the signal, the delay is determined from $|r_{xx}(\Delta t_1)| = 0.1 \cdot r_{xx}(0)$.

- Preferred *direction of a single reflection*: $\pm(55° \pm 20°)$.

- Preferred *amplitude of a single reflection*: $A_1 = \pm 5\,\mathrm{dB}$.

- Preferred *spectrum of a single reflection*: no spectral shaping.

- Preferred *delay time of a second reflection*: $\Delta t_2 = 1.8 \cdot \Delta t_1$.

- Preferred *reverberation time*: $T_{60} = 23 \cdot \Delta t_1$.

These results show that in terms of perception, a preferred pattern of reflections as well as the reverberation time depend decisively on the audio signal. Hence, for different audio signals like classical music, pop music, speech or musical instruments entirely different requirements for early reflections and reverberation time have to be considered.

6.2.2 Gerzon Algorithm

The commonly used method of simulating early reflections is shown in Figs 6.5 and 6.6. The signal is weighted and fed into a system generating early reflections, followed by an addition to the input signal. The first M reflections are implemented by reading samples from a delay line and weighting these samples with a corresponding factor g_i (see Fig. 6.6). The design of a system for simulating early reflections will now be described as proposed by Gerzon [Ger92].

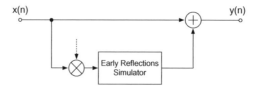

Figure 6.5 Simulation of early reflections.

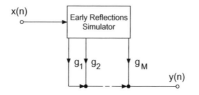

Figure 6.6 Early reflections.

Craven Hypothesis. The Craven hypothesis [Ger92] states that the human perception of the distance to a sound source is evaluated with the help of the amplitude and delay time ratios of the direct signal and early reflections as given by

$$g = \frac{d}{d'}, \tag{6.11}$$

$$T_D = \frac{d' - d}{c} \tag{6.12}$$

$$\Rightarrow d = \frac{cT_D}{g^{-1} - 1}, \tag{6.13}$$

where d is the distance of the source, d' the distance of the image source of the first reflection, g the relative amplitude of the direct signal to first reflection, c the velocity of sound, and T_D the relative delay time of the first reflection to the direct signal.

Without a reflection, human beings are not able to determine the distance d to a sound source. The extended Craven hypothesis includes the absorption coefficient r for determining

$$g = \frac{d}{d'} \exp(-rT_D), \tag{6.14}$$

$$T_D = \frac{d' - d}{c} \tag{6.15}$$

$$\rightarrow d = \frac{cT_D}{g^{-1} \exp(-rT_D) - 1} \tag{6.16}$$

$$\rightarrow g = \frac{\exp(-rT_D)}{1 + cT_D/d}. \tag{6.17}$$

For a given reverberation time T_{60}, the absorption coefficient can be calculated by using $\exp(-rT_{60}) = 1/1000$, that is,

$$r = (\ln 1000)/T_{60}. \tag{6.18}$$

With the relationships (6.15) and (6.17), the parameters for an early reflections simulator as shown in Fig. 6.5 can be determined.

Gerzon's Distance Algorithm. For a system simulating early reflections produced by more than one sound source, Gerzon's distance algorithm can be used [Ger92], where several sound sources are placed at different distances as well as in the stereo position into a stereophonic sound field. An application of this technique is mainly used in multichannel mixing consoles.

By shifting a sound source by $-\delta$ (decrease of relative delay time) it follows that from the relative delay time of the first reflection $T_D - \delta/c = [d' - (d + \delta)]/c$, and the relative amplitude according to (6.17),

$$g_\delta = \frac{1}{1 + \frac{c(T_D - \delta/c)}{d + \delta}} \exp(-r(T_D - \delta/c)) = \left[\frac{d + \delta}{d} \exp(r\delta/c)\right] \frac{\exp(-rT_D)}{1 + cT_D/d}. \qquad (6.19)$$

This results in a delay and a gain factor for the direct signal (see Fig. 6.7) as given by

$$d_2 = d + \delta, \qquad (6.20)$$

$$t_D = \delta/c, \qquad (6.21)$$

$$g_D = \frac{d}{d + \delta} \exp(-r\delta/c). \qquad (6.22)$$

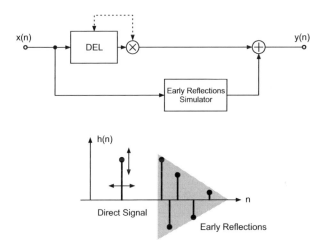

Figure 6.7 Delay and weighting of the direct signal.

By shifting a sound source by $+\delta$ (increase in relative delay time) the relative delay time of the first reflection is $T_D - \delta/c = [d' - (d - \delta)]/c$. As a consequence, the delay and the gain factor for the effect signal (see Fig. 6.8) are given by

$$d_2 = d - \delta, \qquad (6.23)$$

$$t_E = \delta/c, \qquad (6.24)$$

$$g_E = \frac{d}{d + \delta} \exp(-r\delta/c). \qquad (6.25)$$

Using two delay systems in the direct signal as well as in the reflection path, two coupled weighting factors and delay lengths (see Fig. 6.9) can be obtained. For multichannel

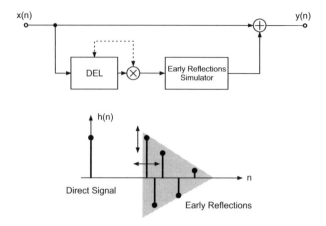

Figure 6.8 Delay and weighting of effect signal.

applications like digital mixing consoles, the scheme in Fig. 6.10 is suggested by Gerzon [Ger92]. Only one system for implementing early reflections is necessary.

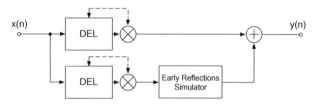

Figure 6.9 Coupled factors and delays.

Stereo Implementation. In many applications, stereo signals have to be processed (see Fig. 6.11). For this purpose, reflections from both sides with positive and negative angles are implemented to avoid stereo displacements. The weighting is done with

$$g_i = \frac{\exp(-rT_i)}{1 + cT_i/d},$$

$$\mathbf{G}_i = g_i \begin{pmatrix} \cos \Theta_i & -\sin \Theta_i \\ \sin \Theta_i & \cos \Theta_i \end{pmatrix}. \tag{6.26}$$

For each reflection, a weighting factor and an angle have to be considered.

Generation of Early Reflection with Increasing Time Density. In [Schr61] it is stated that the time density of reflections increases with the square of time:

$$\text{Number of reflections per second} = (4\pi c^3/V) \cdot t^2. \tag{6.27}$$

After time t_C the reflections have a statistical decay behavior. For a pulse width of Δt, individual reflections overlap after

$$t_C = 5 \cdot 10^{-5}\sqrt{V/\Delta t}. \tag{6.28}$$

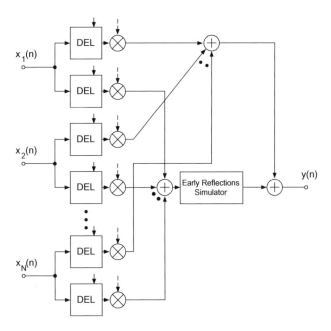

Figure 6.10 Multichannel application.

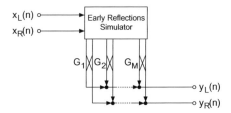

Figure 6.11 Stereo reflections.

To avoid overlap of reflections, Gerzon [Ger92] suggests increasing the density of reflections with t^p (for example, $p = 1$, 0.5 leads to t or $t^{0.5}$). In the interval $(0, 1]$, with initial value x_0 and a number k between 0.5 and 1 the following procedure is performed:

$$y_i = x_0 + ik \pmod 1, \quad i = 0, 1, \ldots, M - 1. \tag{6.29}$$

The numbers y_i in the interval $(0, 1]$ are now transformed to time delays T_i in the interval $[T_{min}, T_{min} + T_{max}]$ by

$$b = T_{min}^{1+p}, \tag{6.30}$$

$$a = (T_{max} + T_{min})^{1+p} - b, \tag{6.31}$$

$$T_i = (ay_i + b)^{1/(1+p)}. \tag{6.32}$$

The increase in the density of reflections is shown by the example in Fig. 6.12.

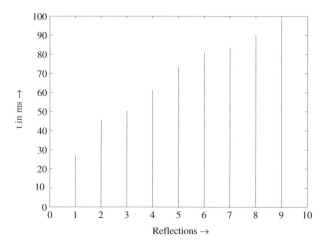

Figure 6.12 Increase in density for nine reflections.

6.3 Subsequent Reverberation

This section deals with techniques for reproducing subsequent reverberation. The first approaches by Schroeder [Schr61, Schr62] and their extension by Moorer [Moo78] will be described. Further developments by Stautner and Puckette [Sta82], Smith [Smi85], Dattarro [Dat97], and Gardner [Gar98] led to general feedback networks [Ger71, Ger76, Jot91, Jot92, Roc95, Roc96, Roc97, Roc02] which have a random impulse response with exponential decay. An extensive discussion on analysis and synthesis parameters of subsequent reverberation can be found in [Ble01]. Apart from the echo density, an important parameter of subsequent reverberation [Cre03] is the quadratic increase in

$$\text{Frequency density} = \frac{4\pi V}{c^3} \cdot f^2 \tag{6.33}$$

with frequency. The following systems exhibit a quadratic increase in echo density and frequency density.

6.3.1 Schroeder Algorithm

The first software implementations of room simulation algorithms were carried out in 1961 by Schroeder. The basis for simulating an impulse response with exponential decay is a recursive comb filter as shown in Fig. 6.13.

The transfer function is given by

$$H(z) = \frac{z^{-M}}{1 - gz^{-M}}, \tag{6.34}$$

$$= \sum_{k=0}^{M-1} \frac{A_k}{z - z_k}, \tag{6.35}$$

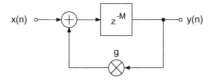

Figure 6.13 Recursive comb filter (g = feedback factor, M = delay length).

with

$$A_k = \frac{z_k}{Mg} \qquad \text{residues,} \tag{6.36}$$

$$z_k = r\, e^{j2\pi k/M} \quad \text{poles,} \tag{6.37}$$

$$r = g^{1/M} \qquad \text{pole radius.} \tag{6.38}$$

With the correspondence of the Z-transform $a/(z-a) \circ\!\!-\!\!\bullet \epsilon(n-1)a^n$ the impulse response is given by

$$H(z) \circ\!\!-\!\!\bullet h(n) = \frac{\epsilon(n-1)}{Mg} \sum_{k=0}^{M-1} z_k^n$$

$$h(n) = \frac{\epsilon(n-1)}{Mg} r^n \sum_{k=0}^{M-1} e^{j\Omega_k n}. \tag{6.39}$$

The complex poles are combined as pairs so that the impulse response can be written as

$$h(n) = \frac{\epsilon(n-1)}{Mg} r^n \sum_{k=1}^{M/2-1} \cos \Omega_k n, \qquad M \text{ even,} \tag{6.40}$$

$$= \frac{\epsilon(n-1)}{Mg} r^n \left[1 + \sum_{k=1}^{M+1/2-1} \cos \Omega_k n \right], \qquad M \text{ uneven.} \tag{6.41}$$

The impulse response is expressed as a summation of cosine oscillations with frequencies Ω_k. These frequencies correspond to the eigenfrequencies of a room. They decay with an exponential envelope r^n, where r is the damping constant (see Fig. 6.15a). The overall impulse response is weighted by $1/Mg$. The frequency response of the comb filter is shown in Fig. 6.15c and is given by

$$|H(e^{j\Omega})| = \sqrt{\frac{1}{1 - 2g\cos(\Omega M) + g^2}}. \tag{6.42}$$

It shows maxima at $\Omega = 2\pi k/M (k = 0, 1, \ldots, M-1)$ of magnitude

$$|H(e^{j\Omega})|_{\text{max}} = \frac{1}{1-g}, \tag{6.43}$$

and minima at $\Omega = (2k+1)\pi/M (k = 0, 1, \ldots, M-1)$ of magnitude

$$|H(e^{j\Omega})|_{\text{min}} = \frac{1}{1+g}. \tag{6.44}$$

Another basis of the Schroeder algorithm is the all-pass filter shown in Fig. 6.14 with transfer function

$$H(z) = \frac{z^{-M} - g}{1 - gz^{-M}} \tag{6.45}$$

$$= \frac{z^{-M}}{1 - gz^{-M}} - \frac{g}{1 - gz^{-M}}. \tag{6.46}$$

From (6.46) it can be seen that the impulse response can also be expressed as a summation of cosine oscillations.

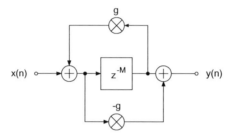

Figure 6.14 All-pass filter ($M =$ delay length).

The impulse responses and frequency responses of a comb filter and an all-pass filter are presented in Fig. 6.15. Both impulse responses show an exponential decay. A sample in the impulse response occurs every M sampling periods. The density of samples in the impulse responses does not increase with time. For the recursive comb filter, spectral shaping due to the maxima at the corresponding poles of the transfer function is observed.

Frequency Density

The frequency density describes the number of eigenfrequencies per hertz and is defined for a comb filter [Jot91] as

$$D_f = M \cdot T_S \quad [1/\text{Hz}]. \tag{6.47}$$

A single comb filter gives M resonances in the interval $[0, 2\pi]$, which are separated by a frequency distance of $\Delta f = f_S/M$. In order to increase the frequency density, a parallel circuit (see Fig. 6.16) of P comb filters is used which leads to

$$H(z) = \sum_{p=1}^{P} \frac{z^{-M_p}}{1 - g_p z^{-M_p}} = \left[\frac{z^{-M_1}}{1 - g_1 z^{-M_1}} + \frac{z^{-M_2}}{1 - g_2 z^{-M_2}} + \cdots \right]. \tag{6.48}$$

The choice of the delay systems [Schr62] is suggested as

$$M_1 : M_P = 1 : 1.5 \tag{6.49}$$

and leads to a frequency density

$$D_f = \sum_{p=1}^{P} M_p \cdot T_S = P \cdot \overline{M} \cdot T_S. \tag{6.50}$$

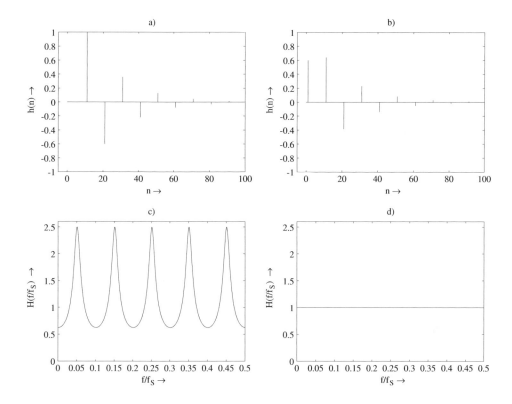

Figure 6.15 (a) Impulse response of a comb filter ($M = 10$, $g = -0.6$). (b) Impulse response of an all-pass filter ($M = 10$, $g = -0.6$). (c) Frequency response of a comb filter. (d) Frequency response of an all-pass filter.

In [Schr62] a necessary frequency density of $D_f = 0.15$ eigenfrequencies per hertz is proposed.

Echo Density

The echo density is the number of reflections per second and is defined for a comb filter [Jot91] as

$$D_t = \frac{1}{M \cdot T_S} \quad [1/s]. \tag{6.51}$$

For a parallel circuit of comb filters, the echo density is given by

$$D_t = \sum_{p=1}^{P} \frac{1}{M_p \cdot T_S} = P \frac{1}{M \cdot T_S}. \tag{6.52}$$

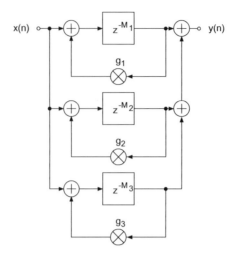

Figure 6.16 Parallel circuit of comb filters.

With (6.50) and (6.52), the number of parallel comb filters and the mean delay length are given by

$$P = \sqrt{D_f \cdot D_t},$$ (6.53)

$$\overline{M}T_S = \sqrt{D_f / D_t}.$$ (6.54)

For a frequency density $D_f = 0.15$ and an echo density $D_t = 1000$ it can be concluded that the number of parallel comb filters is $P = 12$ and the mean delay length is $\overline{M}T_S = 12$ ms. Since the frequency density is proportional to the reverberation time, the number of parallel comb filters has to be increased accordingly.

A further increase in the echo density is achieved by a cascade circuit of P_A all-pass filters (see Fig. 6.17) with transfer function

$$H(z) = \prod_{p=1}^{P_A} \frac{z^{-M_p} - g_p}{1 - g_p z^{-M_p}}.$$ (6.55)

These all-pass sections are connected in series with the parallel circuit of comb filters. For a sufficient echo density, 10000 reflections per second are necessary [Gri89].

Avoiding Unnatural Resonances

Since the impulse response of a single comb filter can be described as a sum of M (delay length) decaying sinusoidal oscillations, the short-time FFT of consecutive parts from this impulse response gives the frequency response shown in Fig. 6.18 in the time-frequency domain. Only the maxima are presented. The parallel circuit of comb filters with the condition (6.49) leads to radii of pole distribution as given by $r_p = g_p^{1/M_p}$ ($p = 1, 2, \ldots, P$). In order to avoid unnatural resonances, the radii of the pole distribution of a parallel circuit

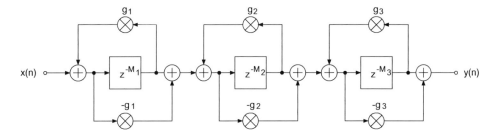

Figure 6.17 Cascade circuit of all-pass filters.

of comb filters must satisfy the condition

$$r_p = \text{const.} = g_p^{1/M_p}, \quad \text{for } p = 1, 2, \ldots, P. \tag{6.56}$$

This leads to the short-time spectra and the pole distribution as shown in Fig. 6.19. Figure 6.20 shows the impulse response and the echogram (logarithmic presentation of the amplitude of the impulse response) of a parallel circuit of comb filters with equal and unequal pole radii. For unequal pole radius, the different decay times of the eigenfrequencies can be seen.

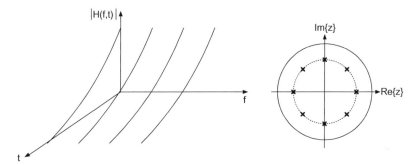

Figure 6.18 Short-time spectra of a comb filter ($M = 8$).

Reverberation Time

The reverberation time of a recursive comb filter can be adjusted with the feedback factor g which describes the ratio

$$g = \frac{h(n)}{h(n - M)} \tag{6.57}$$

of two different nonzero samples of the impulse response separated by M sampling periods. The factor g describes the decay constant per M samples. The decay constant per sampling period can be calculated from the pole radius $r = g^{1/M}$ and is defined as

$$r = \frac{h(n)}{h(n - 1)}. \tag{6.58}$$

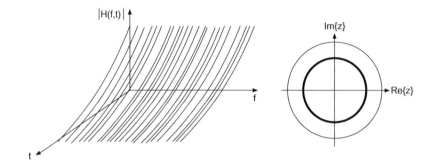

Figure 6.19 Short-time spectra of a parallel circuit of comb filters.

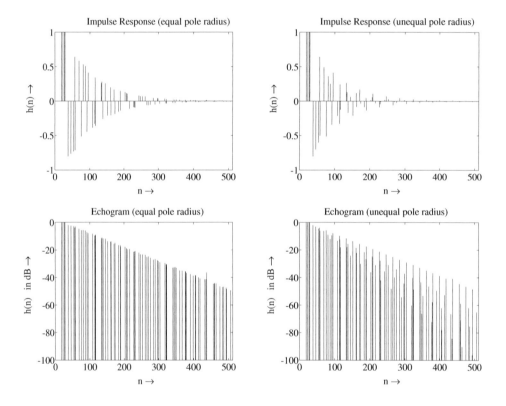

Figure 6.20 Impulse response and echogram.

The relationship between feedback factor g and pole radius r can also be expressed using (6.57) and (6.58) and is given by

$$g = \frac{h(n)}{h(n-M)} = \frac{h(n)}{h(n-1)} \cdot \frac{h(n-1)}{h(n-2)} \cdots \frac{h(n-(M-1))}{h(n-M)} = r \cdot r \cdot r \cdots r = r^M.$$

$$(6.59)$$

With the constant radius $r = g_p^{1/M_p}$ and the logarithmic parameters $R = 20 \log_{10} r$ and $G_p = 20 \log_{10} g_p$, the attenuation per sampling period is given by

$$R = \frac{G_p}{M_p}. \tag{6.60}$$

The reverberation time is defined as decay time of the impulse response to -60 dB. With $-60/T_{60} = R/T_S$, the reverberation time can be written as

$$T_{60} = -60\frac{T_S}{R} = -60\frac{T_S M_p}{G_p} = \frac{3}{\log_{10}|1/g_p|} M_p \cdot T_S. \tag{6.61}$$

The control of reverberation time can either be carried out with the feedback factor g or the delay parameter M. The increase in the reverberation time with factor g is responsible for a pole radius close to the unit circle and, hence, leads to an amplification of maxima of the frequency response (see (6.43)). This leads to a *coloring* of the sound impression. The increase in the delay parameter M, on the other hand, leads to an impulse response whose nonzero samples are far apart from each other, so that individual echoes can be heard. The discrepancy between echo density and frequency density for a given reverberation time can be solved by a sufficient number of parallel comb filters.

Frequency-dependent Reverberation Time

The eigenfrequencies of rooms have a rapid decay for high frequencies. A frequency-dependent reverberation time can be implemented with a low-pass filter

$$H_1(z) = \frac{1}{1 - az^{-1}} \tag{6.62}$$

in the feedback loop of a comb filter. The modified comb filter in Fig. 6.21 has transfer function

$$H(z) = \frac{z^{-M}}{1 - gH_1(z)z^{-M}} \tag{6.63}$$

with the stability criterion

$$\frac{g}{1 - a} < 1. \tag{6.64}$$

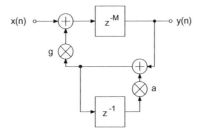

Figure 6.21 Modified low-pass comb filter.

The short-time spectra and the pole distribution of a parallel circuit with low-pass comb filters are presented in Fig. 6.22. Low eigenfrequencies decay more slowly than higher ones. The circular pole distribution becomes an elliptical distribution where the low-frequency poles are moved toward the unit circle.

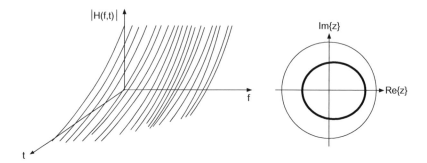

Figure 6.22 Short-time spectra of a parallel circuit of low-pass comb filters.

Stereo Room Simulation

An extension of the Schroeder algorithm was suggested by Moorer [Moo78]. In addition to a parallel circuit of comb filters in series with a cascade of all-pass filters, a pattern of early reflections is generated. Figure 6.23 shows a room simulation system for a stereo signal. The generated room signals $e_L(n)$ and $e_R(n)$ are added to the direct signals $x_L(n)$ and $x_R(n)$. The input of the room simulation is the mono signal $x_M(n) = x_L(n) + x_R(n)$ (sum signal). This mono signal is added to the left and right room signals after going through a delay line DEL1. The total sum of all reflections is fed via another delay line DEL2 to a parallel circuit of comb filters which implements subsequent reverberation. In order to get a high-quality spatial impression, it is necessary to decorrelate the room signals $e_L(n)$ and $e_R(n)$ [Bla74, Bla85]. This can be achieved by taking left and right room signals at different points out of the parallel circuit of comb filters. These room signals are then fed to an all-pass section to increase the echo density.

Besides the described system for stereo room simulation in which the mono signal is processed with a room algorithm, it is also possible to perform complete stereo process-ing of $x_L(n)$ und $x_R(n)$, or to process a mono signal $x_M(n) = x_L(n) + x_R(n)$ and a side (difference) signal $x_S(n) = x_L(n) - x_R(n)$ individually.

6.3.2   General Feedback Systems

Further developments of the comb filter method by Schroeder tried to improve the acoustic quality of reverberation and especially the increase in echo density [Ger71, Ger76, Sta82, Jot91, Jot92, Roc95, Roc96, Roc97a, RS97b]. With respect to [Jot91], the general feedback system in Fig. 6.24 is considered. For the sake of simplicity only three delay systems are shown. The feedback of output signals is carried out with the help of a matrix **A** which feeds back each of the three outputs to the three inputs.

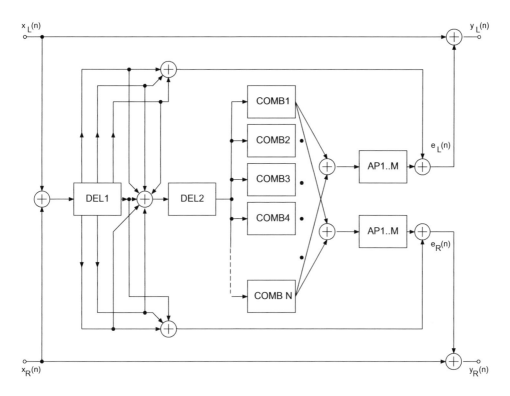

Figure 6.23 Stereo room simulation.

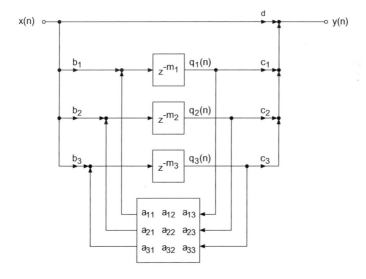

Figure 6.24 General feedback system.

In general, for N delay systems we can write

$$y(n) = \sum_{i=1}^{N} c_i q_i(n) + d x(n), \tag{6.65}$$

$$q_j(n + m_j) = \sum_{i=1}^{N} a_{ij} q_i(n) + b_j x(n), \quad 1 \leq j \leq N. \tag{6.66}$$

The Z-transform leads to

$$Y(z) = \mathbf{c}^T \mathbf{Q}(z) + d \cdot X(z), \tag{6.67}$$

$$\mathbf{D}(z) \cdot \mathbf{Q}(z) = \mathbf{A} \cdot \mathbf{Q}(z) + \mathbf{b} \cdot X(z)$$
$$\rightarrow \mathbf{Q}(z) = [\mathbf{D}(z) - \mathbf{A}]^{-1} \mathbf{b} \cdot X(z), \tag{6.68}$$

with

$$\mathbf{Q}(z) = \begin{bmatrix} Q_1(z) \\ \vdots \\ Q_N(z) \end{bmatrix}, \quad \mathbf{b} = \begin{bmatrix} b_1 \\ \vdots \\ b_N \end{bmatrix}, \quad \mathbf{c} = \begin{bmatrix} c_1 \\ \vdots \\ c_N \end{bmatrix} \tag{6.69}$$

and the diagonal delay matrix

$$\mathbf{D}(z) = \mathrm{diag}[z^{-m_1} \cdots z^{-m_N}]. \tag{6.70}$$

With (6.68) the Z-transform of the output is given by

$$Y(z) = \mathbf{c}^T [\mathbf{D}(z) - \mathbf{A}]^{-1} \mathbf{b} \cdot X(z) + d \cdot X(z) \tag{6.71}$$

and the transfer function by

$$H(z) = \mathbf{c}^T [\mathbf{D}(z) - \mathbf{A}]^{-1} \mathbf{b} + d. \tag{6.72}$$

The system is stable if the feedback matrix $\mathbf{A}$ can be expressed as a product of a unitary matrix $\mathbf{U}$ ($\mathbf{U}^{-1} = \overline{\mathbf{U}}^T$) and a diagonal matrix with $g_{ii} < 1$ (derivation in [Sta82]). Figure 6.25 shows a general feedback system with input vector $\mathbf{X}(z)$, output vector $\mathbf{Y}(z)$, a diagonal matrix $\mathbf{D}(z)$ consisting of purely delay systems z^{-m_i}, and a feedback matrix $\mathbf{A}$. This feedback matrix consists of an orthogonal matrix $\mathbf{U}$ multiplied by the matrix $\mathbf{G}$ which results in a weighting of the feedback matrix $\mathbf{A}$.

If an orthogonal matrix $\mathbf{U}$ is chosen and the weighting matrix is equal to the unit matrix $\mathbf{G} = \mathbf{I}$, the system in Fig. 6.25 implements a white-noise random signal with Gaussian distribution when a pulse excitation is applied to the input. The time density of this signal slowly increases with time. If the diagonal elements of the weighting matrix $\mathbf{G}$ are less than one, a random signal with exponential amplitude decay results. With the help of the weighting matrix $\mathbf{G}$, the reverberation time can be adjusted. Such a feedback system performs the convolution of an audio input signal with an impulse response of exponential decay.

The effect of the orthogonal matrix $\mathbf{U}$ on the subjective sound perception of subsequent reverberation is of particular interest. A relationship between the distribution of the eigenvalues of the matrix $\mathbf{U}$ on the unit circle and the poles of the system transfer function cannot

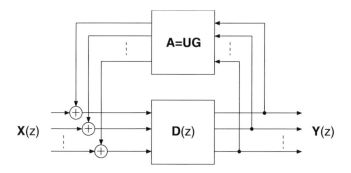

Figure 6.25 Feedback system.

be described analytically, owing to the high order of the feedback system. In [Her94], it is shown experimentally that the distribution of eigenvalues within the right-hand or left-hand complex plane produces a uniform distribution of poles of the system transfer function. Such a feedback matrix leads to an acoustically improved reverberation. The echo density rapidly increases to the maximum value of one sample per sampling period for a uniform distribution of eigenvalues. Besides the feedback matrix, additional digital filtering is necessary for spectrally shaping the subsequent reverberation and for implementing frequency-dependent decay times (see [Jot91]). The following example illustrates the increase of the echo density.

Example: First, a system with only one feedback path per comb filter is considered. The feedback matrix is then given by

$$\mathbf{A} = \frac{g}{\sqrt{2}} \mathbf{I}. \tag{6.73}$$

Figure 6.26 shows the impulse response and the amplitude frequency response.

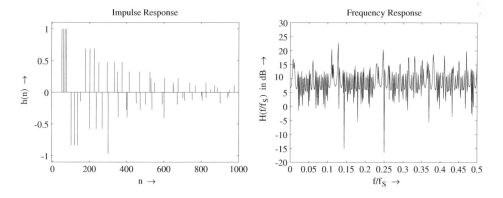

Figure 6.26 Impulse response and frequency response of a 4-delay system with a unit matrix as unitary feedback matrix ($g = 0.83$).

With the feedback matrix

$$\mathbf{A} = \frac{g}{\sqrt{2}} \begin{bmatrix} 0 & 1 & 1 & 0 \\ -1 & 0 & 0 & -1 \\ 1 & 0 & 0 & -1 \\ 0 & 1 & -1 & 0 \end{bmatrix} \qquad (6.74)$$

from [Sta82], the impulse response and the corresponding frequency response shown in Fig. 6.27 are obtained. In contrast to Fig. 6.26, an increase in the echo density of the impulse response is observed.

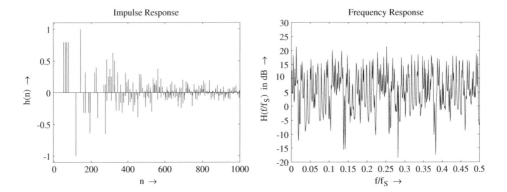

Figure 6.27 Impulse response and frequency response of a 4-delay system with unitary feedback matrix ($g = 0.63$).

6.3.3 Feedback All-pass Systems

In addition to the general feedback systems, simple delay systems with feedback have been used for room simulators (see Fig. 6.28). Theses simulators are based on a delay line, where single delays are fed back with L feedback coefficients to the input. The sum of the input signal and feedback signal is low-pass filtered or spectrally weighted by a low-frequency shelving filter and is then put to the delay line again. The first N reflections are extracted out of the delay line according to the reflection pattern of the simulated room. They are weighted and added to the output signal. The mixing between the direct signal and the room signal is adjusted by the factor g_{MIX}. The inner system can be described by a rational transfer function $H(z) = Y(z)/X(z)$. In order to avoid a low frequency density the feedback delay lengths can be made time-variant [Gri89, Gri91].

Increasing the echo density can be achieved by replacing the delays z^{-M_i} by frequency dependent all-pass systems $A(z^{-M_i})$. This extension was first proposed by Gardner in [Gar92, Gar98]. In addition to the replacement of $z^{-M_i} \to A(z^{-M_i})$, the all-pass systems can be extended by *embedded all-pass systems* [Gar92]. Figure 6.29 shows an all-pass system (Fig. 6.29a) where the delay z^{-M} is replaced by a further all-pass and a unit delay z^{-1} (Fig. 6.29b). The integration of a unit delay avoids delay-free loops. In Fig. 6.29c the inner all-pass is replaced by a cascade of two all-pass systems and a further delay z^{-M_3}.

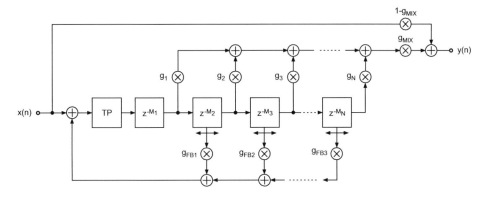

Figure 6.28 Room simulation with delay line and forward and backward coefficients.

The resulting system is again an all-pass system [Gar92, Gar98]. A further modification of the general all-pass system is shown in Fig. 6.29d [Dat97, Vää97, Dah00]. Here, a delay z^{-M} followed by a low-pass and a weighting coefficient is used. The resulting system is called an *absorbent all-pass system*. With these embedded all-pass systems the room simulator shown in Fig. 6.28 is extended to a feedback all-pass system which is shown in Fig. 6.30 [Gar92,Gar98]. The feedback is performed by a low-pass filter and a feedback coefficient g, which adjusts the decay behavior. The extension to a stereo room simulator is described in [Dat97, Dah00] and is depicted in Fig. 6.31 [Dah00]. The cascaded all-pass systems $A_i(z)$ in the left and right channel can be a combination of embedded and absorbent all-pass systems. Both output signals of the all-pass chains are fed back to the input and added. In front of both all-pass chains a coupling of both channels with a weighted sum and difference is performed. The setup and parameters of such a system are discussed in [Dah00]. A precise adjustment of reverberation time and control of echo density can be achieved by the feedback coefficients of the all-passes. The frequency density is controlled by the scaling of the delay lengths of the inner all-pass systems.

6.4 Approximation of Room Impulse Responses

In contrast to the systems for simulation of room impulse responses discussed up to this point, a method is now presented that measures and approximates the room impulse response in one step [Zöl90b, Sch92, Sch93] (see Fig. 6.32). Moreover, it leads to a parametric representation of the room impulse response. Since the decay times of room impulse responses decrease for high frequencies, use is made of multirate signal processing.

The analog system that is to be measured and approximated is excited with a binary pseudo-random sequence $x(n)$ via a DA converter. The resulting room signal gives a digital sequence $y(n)$ after AD conversion. The discrete-time sequence $y(n)$ and the pseudo-random sequence $x(n)$ are each decomposed by an analysis filter bank into subband signals $y_1, \ldots, y_P$ and $x_1, \ldots, x_P$ respectively. The sampling rate is reduced in accordance with the bandwidth of the signals. The subband signals $y_1, \ldots, y_P$ are approximated by adjusting the subband systems $H_1(z) = A_1(z)/B_1(z), \ldots, H_P(z) = A_P(z)/B_P(z)$. The outputs

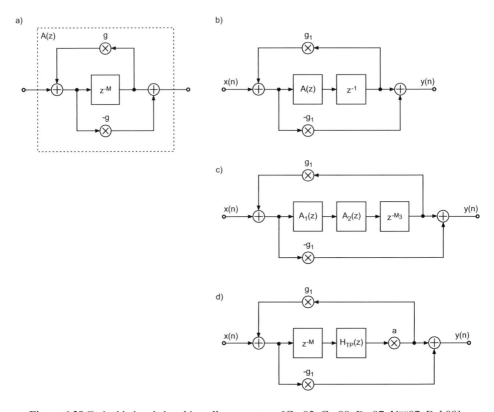

Figure 6.29 Embedded and absorbing all-pass system [Gar92, Gar98, Dat97, Vää97, Dah00].

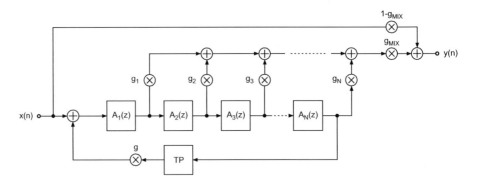

Figure 6.30 Room simulator with embedded all-pass systems [Gar92,Gar98].

$\hat{y}_1, \ldots, \hat{y}_P$ of these subband systems give an approximation of the measured subband signals. With this procedure, the impulse response is given in parametric form (subband parameters) and can be directly simulated in the digital domain.

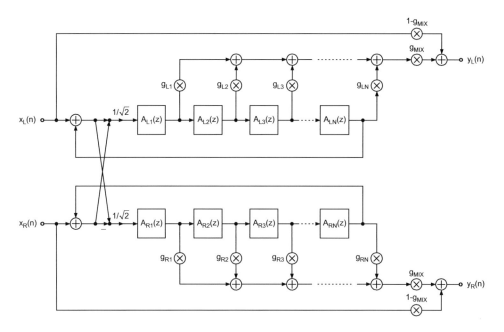

Figure 6.31 Stereo room simulator with absorbent all-pass systems [Dah00].

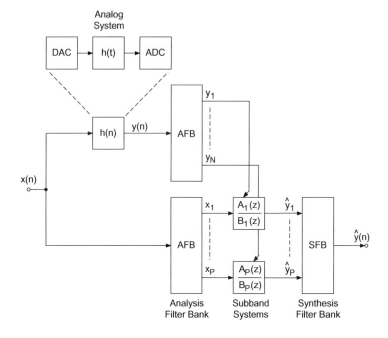

Figure 6.32 System measuring and approximating room impulse responses.

By suitably adjusting the analysis filter bank [Sch94], the subband impulse responses are obtained directly from the cross-correlation function

$$h_i \approx r_{x_i y_i}. \tag{6.75}$$

The subband impulse responses are approximated by a nonrecursive filter and a recursive comb filter. The cascade of both filters leads to the transfer function

$$H_i(z) = \frac{b_0 + \cdots + b_{M_i} z^{-M_i}}{1 - g_i z^{-N_i}} = \sum_{n_i=0}^{\infty} h_i(n_i) z^{-n_i}, \tag{6.76}$$

which is set equal to the impulse response in subband i. Multiplying both sides of (6.76) by the denominator $1 - g_i z^{-N_i}$ gives

$$b_0 + \cdots + b_{M_i} z^{-M_i} = \left(\sum_{n_i=0}^{\infty} h_i(n_i) z^{-n_i} \right) (1 - g_i z^{-N_i}). \tag{6.77}$$

Truncating the impulse response of each subband to K samples and comparing the coefficients of powers of z on both sides of the equation, the following set of equations is obtained:

$$
\begin{bmatrix}
b_0 \\
b_1 \\
\vdots \\
b_M \\
0 \\
\vdots \\
0
\end{bmatrix}
=
\begin{bmatrix}
h_0 & 0 & 0 & \cdots & 0 \\
h_1 & h_0 & 0 & \cdots & 0 \\
\vdots & \vdots & \vdots & & \vdots \\
h_M & h_{M-1} & h_{M-2} & \cdots & h_{M-N} \\
h_{M+1} & h_M & h_{M-1} & \cdots & h_{M-N+1} \\
\vdots & \vdots & \vdots & & \vdots \\
h_K & h_{K-1} & h_{K-2} & \cdots & h_{K-N}
\end{bmatrix}
\begin{bmatrix}
1 \\
0 \\
\vdots \\
-g
\end{bmatrix}. \tag{6.78}
$$

The coefficients $b_0, \ldots, b_M$ and g in the above equation are determined in two steps. First, the coefficient g of the comb filter is calculated from the exponentially decaying envelope of the measured subband impulse response. The vector $[1, 0, \ldots, g]^T$ is then used to determine the coefficients $[b_0, b_1, \ldots, b_M]^T$.

For the calculation of the coefficient g, we start with the impulse response of the comb filter $H(z) = 1/(1 - gz^{-N})$ given by

$$h(l = Nn) = g^l. \tag{6.79}$$

We further make use of the *integrated impulse response*

$$h_e(k) = \sum_{n=k}^{\infty} h(n)^2 \tag{6.80}$$

defined in [Schr65]. This describes the rest energy of the impulse response at time k. By taking the logarithm of $h_e(k)$, a straight line over time index k is obtained. From the slope of the straight line we use

$$\ln g = N \cdot \frac{\ln h_e(n_1) - \ln h_e(n_2)}{n_1 - n_2}, \quad n_1 < n_2, \tag{6.81}$$

to determine the coefficient g [Sch94]. For $M = N$, the coefficients in (6.78) of the numerator polynomial are obtained directly from the impulse response

$$b_n = h_n, \quad n = 0, 1, \ldots, M - 1,$$
$$b_M = h_M - g h_0. \tag{6.82}$$

Hence, the numerator polynomial of (6.76) is a direct reproduction of the first M samples of the impulse response (see Fig. 6.33). The denominator polynomial approximates the further exponentially decaying impulse response. This method is applied to each subband. The implementation complexity can be reduced by a factor of 10 compared with the direct implementation of the broad-band impulse response [Sch94]. However, owing to the group delay caused by the filter bank, this method is not so suitable for real-time applications.

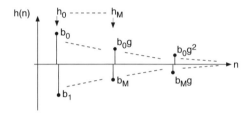

Figure 6.33 Determining model parameters from the measured impulse response.

6.5 Java Applet – Fast Convolution

The applet shown in Fig. 6.34 demonstrates audio effects resulting from a fast convolution algorithm. It is designed for a first insight into the perceptual effects of convolving an impulse response with an audio signal.

The applet generates an impulse response by modulating the amplitude of a random signal. The graphical interface presents the curve of the amplitude modulation, which can be manipulated with three control points. Two control points are used for the initial behavior of the amplitude modulation. The third control point is used for the exponential decay of the impulse response. You can choose between two predefined audio files from our web server (*audio1.wav* or *audio2.wav*) or your own local wav file to be processed [Gui05].

6.6 Exercises

1. Room Impulse Responses

1. How can we measure a room impulse response?

2. What kind of test signal is necessary?

3. How does the length of the impulse response affect the length of the test signal?

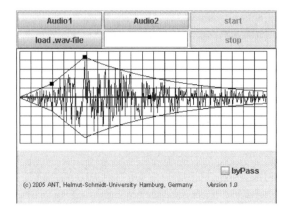

Figure 6.34 Java applet – fast convolution.

2. First Reflections

For a given sound (voice sound) calculate the delay time of a single first reflection. Write a Matlab program for the following computations.

1. How do we choose this delay time? What coefficient should be used for it?

2. Write an algorithm which performs the convolution of the input mono signal with two impulse responses which simulate a reflection to the left output $y_L(n)$ and a second reflection to the right output $y_R(n)$. Check the results by listening to the output sound.

3. Improve your algorithm to simulate two reflections which can be positioned at any angle inside the stereo mix.

3. Comb and All-pass Filters

1. **Comb Filters:** Based on the Schroeder algorithm, draw a signal flow graph for a comb filter consisting of a single delay line of M samples with a feedback loop containing an attenuation factor g.

   (a) Derive the transfer function of the comb filter.

   (b) Now the attenuation factor g is in the feed-forward path and in the feedback loop no attenuation is applied. Why can we consider the impulse response of this model to be similar to the previous one?

   (c) In both cases how should we choose the gain factor? What will happen if we do otherwise?

   (d) Calculate the reverberation time of the comb filter for $f_S = 44.1\,\text{kHz}$, $M = 8$ and g specified previously.

   (e) Write down what you know about the filter coefficients, plot the pole/zero locations and the frequency response of the filter

2. **All-pass Filters:** Realize an all-pass structure as suggested by Schroeder.

   (a) Why can we expect a better result with an all-pass filter than with comb filter? Write a Matlab function for a comb and all-pass filter with $M = 8, 16$.

   (b) Derive the transfer function and show the pole/zero locations, the impulse response, the magnitude and phase responses.

   (c) Perform the filtering of an audio signal with the two filters and estimate the delay length M which leads to a perception of a room impression.

4. Feedback Delay Networks

Write a Matlab program which realizes a feedback delay network.

   1. What is the reason for a unitary feedback matrix?

   2. What is the advantage of using a unitary circulant feedback matrix?

   3. How do you control the reverberation time?

References

[All79]   J. B. Allen, D. A. Berkeley: *Image Method for Efficient Simulating Small Room Acoustics*, J. Acoust. Soc. Am., Vol. 65, No. 4, pp. 943–950, 1979.

[And90]   Y. Ando: *Concert Hall Acoustics*, Springer-Verlag, Berlin, 1990.

[Bro01]   S. Browne: *Hybrid Reverberation Algorithm Using Truncated Impulse Response Convolution and Recursive Filtering*, MSc Thesis, University of Miami, Coral Gables, FL, June 2001.

[Bar71]   M. Barron: *The Subjective Effects of First Reflections in Concert Halls – The Need for Lateral Reflections*, J. Sound and Vibration, Vol. 15, pp. 475–494, 1971.

[Bar81]   M. Barron, A. H. Marschall: *Spatial Impression Due to Early Lateral Reflections in Concert Halls: The Derivation of a Physical Measure*, J. Sound and Vibration, Vol. 77, pp. 211–232, 1981.

[Bel99]   F. A. Beltran, J. R. Beltran, N. Holzem, A. Gogu: *Matlab Implementation of Reverberation Algorithms*, Proc. Workshop on Digital Audio Effects (DAFx-99), pp. 91–96, Trondheim, 1999.

[Bla74]   J. Blauert: *Räumliches Hören*, S. Hirzel Verlag, Stuttgart, 1974.

[Bla85]   J. Blauert: *Räumliches Hören*, Nachschrift-Neue Ergebnisse und Trends seit 1972, S. Hirzel Verlag, Stuttgart, 1985.

[Ble01]   B. Blesser: *An Interdisciplinary Synthesis of Reverberation Viewpoints*, J. Audio Eng. Soc., Vol. 49, pp. 867–903, October 2001.

[Cre78]   L. Cremer, H. A. Müller: *Die wissenschaftlichen Grundlagen der Raumakustik*, Vols. 1 and 2, S. Hirzel Verlag, Stuttgart, 1976/78.

[Cre03]   L. Cremer, M. Möser: *Technische Akustik*, Springer-Verlag, Berlin, 2003.

[Dah00]   L. Dahl, J.-M. Jot: *A Reverberator Based on Absorbent All-Pass Filters*, in Proc. of the COST G-6 Conference on Digital Audio Effects (DAFX-00), Verona, December 2000.

[Dat97]   J. Dattorro: *Effect Design – Part 1: Reverberator and Other Filters*, J. Audio Eng. Soc., Vol. 45, pp. 660–684, September 1997.

[Ege96]   G. P. M. Egelmeers, P. C. W. Sommen: *A New Method for Efficient Convolution in Frequency Domain by Nonuniform Partitioning for Adaptive Filtering*, IEEE Trans. Signal Processing, Vol. 44, pp. 3123–3192, December 1996.

[Far00]   A. Farina: *Simultaneous Measurement of Impulse Response and Distortion with a Swept-Sine Technique*, Proc. 108th AES Convention, Paris, February 2000.

[Fre00]   J. Frenette: *Reducing Artificial Reverberation Algorithm Requirements Using Time-Varying Feddback Delay Networks*, MSc thesis, University of Miami, Coral Gables, FL, December 2000.

[Gar95]   W. G. Gardner: *Efficient Convolution Without Input-Output Delay*, J. Audio Eng. Soc., Vol. 43, pp. 127–136, 1995.

[Gar98]   W. G. Gardner: *Reverberation Algorithms*, in M. Kahrs and K. Brandenburg (Ed.), Applications of Digital Signal Processing to Audio and Acoustics, Kluwer Academic Publishers, Boston, pp. 85–131, 1998.

[Ger71]   M. A. Gerzon: *Synthetic Stereo Reverberation*, Studio Sound, No. 13, pp. 632–635, 1971 and No. 14, pp. 24–28, 1972.

[Ger76]   M. A. Gerzon: *Unitary (Energy-Preserving) Multichannel Networks with Feedback*, Electronics Letters, Vol. 12, No. 11, pp. 278–279, 1976.

[Ger92]   M. A. Gerzon: *The Design of Distance Panpots*, Proc. 92nd AES Convention, Preprint No. 3308, Vienna, 1992.

[Gri89]   D. Griesinger: *Practical Processors and Programs for Digital Reverberation*, Proc. AES 7th Int. Conf., pp. 187–195, Toronto, 1989.

[Gri91]   D. Griesinger: *Improving Room Acoustics through Time-Variant Synthetic Reverberation*, in Proc. 90th Conv. Audio Eng. Soc., Preprint 3014, February 1991.

[Gui05]   M. Guillemard, C. Ruwwe, U. Zölzer: *J-DAFx – Digital Audio Effects in Java*, Proc. 8th Int. Conference on Digital Audio Effects (DAFx-05), pp. 161–166, Madrid, 2005.

[Her94]   T. Hertz: *Implementierung und Untersuchung von Rückkopplungssystemen zur digitalen Raumsimulation*, Diplomarbeit, TU Hamburg-Harburg, 1994.

[Iid95]   K. Iida, K. Mizushima, Y. Takagi, T. Suguta: *A New Method of Generating Artificial Reverberant Sound*, Proc. 99th AES Convention, Preprint No. 4109, October 1995.

[Joh00]   M. Joho, G. S. Moschytz: *Connecting Partitioned Frequency-Domain Filters in Parallel or in Cascade*, IEEE Trans. CAS-II: Analog and Digital Signal Processing, Vol. 47, No. 8, pp. 685–698, August 2000.

[Jot91]   J. M. Jot, A. Chaigne: *Digital Delay Networks for Designing Artificial Reverberators*, Proc. 94th AES Convention, Preprint No. 3030, Berlin, 1991.

[Jot92]   J. M. Jot: *An Analysis/Synthesis Approach to Real-Time Artificial Reverberation*, Proc. ICASSP-92, pp. 221–224, San Francisco, 1992.

[Ken95a]   G. S. Kendall: *A 3-D Sound Primer: Directional Hearing and Stereo Reproduction*, Computer Music J., Vol. 19, No. 4, pp. 23–46, Winter 1995.

[Ken95b]   G. S. Kendall. *The Decorrelation of Audio Signals and Its Impact on Spatial Imagery*, Computer Music J., Vol. 19, No. 4, pp. 71–87, Winter 1995.

[Kut91]   H. Kuttruff: *Room Acoustics*, 3rd edn, Elsevier Applied Sciences, London, 1991.

[Lee03a]   W.-C. Lee, C.-M. Liu, C.-H. Yang, J.-I. Guo, *Perceptual Convolution for Reverberation*, Proc. 115th Convention 2003, Preprint No. 5992, October 2003

[Lee03b]   W.-C. Lee, C.-M. Liu, C.-H. Yang, J.-I. Guo, *Fast Perceptual Convolution for Reverberation*, Proc. of the 6th Int. Conference on Digital Audio Effects (DAFX-03), London, September 2003.

[Lok01]   T. Lokki, J. Hiipakka: *A Time-Variant Reverberation Algorithm for Reverberation Enhancement Systems*, Proc. of the COST G-6 Conference on Digital Audio Effects (DAFX-01), Limerick, December 2001.

[Moo78]   J. A. Moorer: *About This Reverberation Business*, Computer Music J., Vol. 3, No. 2, pp. 13–28, 1978.

[Mac76]   F. J. MacWilliams, N. J. A. Sloane: *Pseudo-Random Sequences and Arrays*, IEEE Proc., Vol. 64, pp. 1715–1729, 1976.

[Mül01]   S. Müller, P. Massarani: *Transfer-Function Measurement with Sweeps*, J. Audio Eng. Soc., Vol. 49, pp. 443–471, 2001.

[Pii98]   E. Piirilä, T. Lokki, V. Välimäki, *Digital Signal Processing Techniques for Non-exponentially Decaying Reverberation*, Proc. 1st COST-G6 Workshop on Digital Audio Effects (DAFX98), Barcelona, 1998.

[Soo90]   J. S. Soo, K. K. Pang: *Multidelay Block Frequency Domain Adaptive Filter*, IEEE Trans. ASSP, Vol. 38, No. 2, pp. 373–376, February 1990.

[Rei95]   A. J. Reijen, J. J. Sonke, D. de Vries: *New Developments in Electro-Acoustic Reverberation Technology*, Proc. 98th AES Convention 1995, Preprint No. 3978, February 1995.

[Roc95]   D. Rocchesso: *The Ball within the Box: A Sound-Processing Metaphor*, Computer Music J., Vol. 19, No. 4, pp. 47–57, Winter 1995.

[Roc96]   D. Rocchesso: *Strutture ed Algoritmi per l'Elaborazione del Suono basati su Reti di Linee di Ritardo Interconnesse*, PhD thesis, University of Padua, February 1996.

[Roc97a]   D. Rocchesso: *Maximally-Diffusive Yet Efficient Feedback Delay Networks for Artificial Reverberation*, IEEE Signal Processing Letters, Vol. 4, No. 9, pp. 252–255, September 1997.

[RS97b]   D. Rocchesso, J. O. Smith: *Circulant and Elliptic Feedback Delay Net- works for Artificial Reverberation*, IEEE Trans. Speech and Audio Processing, Vol. 5, No. 1, pp. 51–63, January 1997.

[Roc02]   D. Rocchesso: *Spatial Effects*, in U. Zölzer (Ed.), DAFX – Digital Audio Effects, John Wiley & Sons, Ltd, Chichester, pp. 137–200, 2002.

[Sch92]   M. Schönle, U. Zölzer, N. Fliege: *Modeling of Room Impulse Responses by Multirate Systems*, Proc. 93rd AES Convention, Preprint No. 3447, San Francisco, 1992.

[Sch93]   M. Schönle, N. J. Fliege, U. Zölzer: *Parametric Approximation of Room Impulse Responses by Multirate Systems*, Proc. ICASSP-93, Vol. 1, pp. 153–156, 1993.

[Sch94]   M. Schönle: *Wavelet-Analyse und parametrische Approximation von Raumimpulsantworten*, Dissertation, TU Hamburg-Harburg, 1994.

[Schr61]   M. R. Schroeder, B. F. Logan: *Colorless Artificial Reverberation*, J. Audio Eng. Soc., Vol. 9, pp. 192–197, 1961.

[Schr62]   M. R. Schroeder: *Natural Sounding Artificial Reverberation*, J. Audio Eng. Soc., Vol. 10, pp. 219–223, 1962.

[Schr65]   M. R. Schroeder: *New Method of Measuring Reverberation Time*, J. Acoust. Soc. Am., Vol. 37, pp. 409–412, 1965.

[Schr70]   M. R. Schroeder: *Digital Simulation of Sound Transmission in Reverberant Spaces*, J. Acoust. Soc. Am., Vol. 47, No. 2, pp. 424–431, 1970.

[Schr87]   M. R. Schroeder: *Statistical Parameters of the Frequency Response Curves of Large Rooms*, J. Audio Eng. Soc., Vol. 35, pp. 299–305, 1987.

[Smi85]   J. O. Smith: *A New Approach to Digital Reverberation Using Closed Waveguide Networks*, Proc. International Computer Music Conference, pp. 47–53, Vancouver, 1985.

[Sta02]   G. B. Stan, J. J. Embrechts, D. Archambeau: *Comparison of Different Impulse Response Measurement Techniques*, J. Audio Eng. Soc., Vol. 50, pp. 249–262, 2002.

[Sta82]   J. Stautner, M. Puckette: *Designing Multi-channel Reverberators*, Computer Music J., Vol. 6, No. 1, pp. 56–65, 1982.

[Vää97]   R. Väänänen, V. Välimäki, J. Huopaniemi: *Efficient and Parametric Reverberator for Room Acoustics Modeling*, Proc. Int. Computer Music Conf. (ICMC'97), pp. 200–203, Thessaloniki, September 1997.

[Zöl90]   U. Zölzer, N. J. Fliege, M. Schönle, M. Schusdziarra: *Multirate Digital Reverberation System*, Proc. 89th AES

Convention, Preprint No. 2968, Los Angeles, 1990.

Chapter 7

Dynamic Range Control

The dynamic range of a signal is defined as the logarithmic ratio of maximum to minimum signal amplitude and is given in decibels. The dynamic range of an audio signal lies between 40 and 120 dB. The combination of level measurement and adaptive signal level adjustment is called *dynamic range control*. Dynamic range control of audio signals is used in many applications to match the dynamic behavior of the audio signal to different requirements. While recording, dynamic range control protects the AD converter from overload or is employed in the signal path to optimally use the full amplitude range of a recording system. For suppressing low-level noise, so-called noise gates are used so that the audio signal is passed through only from a certain level onwards. While reproducing music and speech in a car, shopping center, restaurant or disco the dynamics have to match the special noise characteristics of the environment. Therefore the signal level is measured from the audio signal and a control signal is derived which then changes the signal level to control the loudness of the audio signal. This loudness control is adaptive to the input level.

7.1 Basics

Figure 7.1 shows a block diagram of a system for dynamic range control. After measuring the input level $X_{\mathrm{dB}}(n)$, the output level $Y_{\mathrm{dB}}(n)$ is affected by multiplying the delayed input signal $x(n)$ by a factor $g(n)$ according to

$$y(n) = g(n) \cdot x(n - D). \tag{7.1}$$

The delay of the signal $x(n)$ compared with the control signal $g(n)$ allows predictive control of the output signal level. This multiplicative weighting is carried out with corresponding attack and release time. Multiplication leads, in terms of a logarithmic level representation of the corresponding signals, to the addition of the weighting level $G_{\mathrm{dB}}(n)$ to the input level $X_{\mathrm{dB}}(n)$, giving the output level

$$Y_{\mathrm{dB}}(n) = X_{\mathrm{dB}}(n) + G_{\mathrm{dB}}(n). \tag{7.2}$$

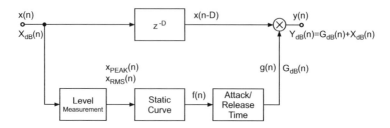

Figure 7.1 System for dynamic range control.

7.2   Static Curve

The relationship between input level and weighting level is defined by a static level curve $G_{dB}(n) = f(X_{dB}(n))$. An example of such a static curve is given in Fig. 7.2. Here, the output level and the weighting level are given as functions of the input level.

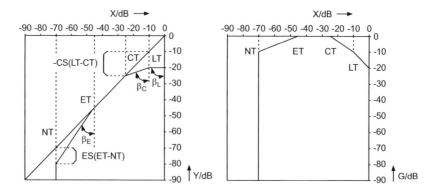

Figure 7.2 Static curve with the parameters (LT $=$ *limiter threshold*, CT $=$ *compressor threshold*, ET $=$ *expander threshold* and NT $=$ *noise gate threshold*).

With the help of a limiter, the output level is limited when the input level exceeds the limiter threshold LT. All input levels above this threshold lead to a constant output level. The compressor maps a change of input level to a certain smaller change of output level. In contrast to a limiter, the compressor increases the loudness of the audio signal. The expander increases changes in the input level to larger changes in the output level. With this, an increase in the dynamics for low levels is achieved. The noise gate is used to suppress low-level signals, for noise reduction and also for sound effects like truncating the decay of room reverberation. Every threshold used in particular parts of the static curve is defined as the lower limit for limiter and compressor and upper limit for expander and noise gate.

In the logarithmic representation of the static curve the compression factor R (*ratio*) is defined as the ratio of the input level change ΔP_I to the output level change ΔP_O:

$$R = \frac{\Delta P_I}{\Delta P_O}. \qquad (7.3)$$

With the help of Fig. 7.3 the straight line equation $Y_{dB}(n) = CT + R^{-1}(X_{dB}(n) - CT)$ and the compression factor

$$R = \frac{X_{dB}(n) - CT}{Y_{dB}(n) - CT} = \tan \beta_C \tag{7.4}$$

are obtained, where the angle β is defined as shown in Fig. 7.2. The relationship between the *ratio* R and the *slope* S can also be derived from Fig. 7.3 and is expressed as

$$S = 1 - \frac{1}{R}, \tag{7.5}$$

$$R = \frac{1}{1 - S}. \tag{7.6}$$

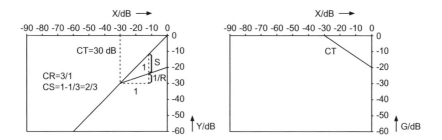

Figure 7.3 Compressor curve (compressor ratio *CR*/slope *CS*).

Typical compression factors are

$$\begin{aligned}
R &= \infty, & \text{limiter,} \\
R &> 1, & \text{compressor } (CR\text{: compressor ratio}), \\
0 < R &< 1, & \text{expander } (ER\text{: expander ratio}), \\
R &= 0, & \text{noise gate.}
\end{aligned} \tag{7.7}$$

The transition from logarithmic to linear representation leads, from (7.4), to

$$R = \frac{\log_{10} \frac{\hat{x}(n)}{c_T}}{\log_{10} \frac{\hat{y}(n)}{c_T}}, \tag{7.8}$$

where $\hat{x}(n)$ and $\hat{y}(n)$ are the linear levels and c_T denotes the linear compressor threshold. Rewriting (7.8) gives the linear output level

$$\frac{\hat{y}(n)}{c_T} = 10^{1/R \log_{10}(\hat{x}(n)/c_T)} = \left(\frac{\hat{x}(n)}{c_T}\right)^{1/R}$$

$$\hat{y}(n) = c_T^{1-1/R} \cdot \hat{x}^{1/R}(n) \tag{7.9}$$

as a function of input level. The control factor $g(n)$ can be calculated by the quotient

$$g(n) = \frac{\hat{y}(n)}{\hat{x}(n)} = \left(\frac{\hat{x}(n)}{c_T}\right)^{1/R-1}. \tag{7.10}$$

With the help of tables and interpolation methods, it is possible to determine the control factor without taking logarithms and antilogarithms. The implementation described as follows, however, makes use of the logarithm of the input level and calculates the control level $G_{dB}(n)$ with the help of the straight line equation. The antilogarithm leads to the value $f(n)$ which gives the control factor $g(n)$ with corresponding attack and release time (see Fig. 7.1).

7.3 Dynamic Behavior

Besides the static curve of dynamic range control, the dynamic behavior in terms of attack and release times plays a significant role in sound quality. The rapidity of dynamic range control depends also on the measurement of PEAK and RMS values [McN84, Sti86].

7.3.1 Level Measurement

Level measurements [McN84] can be made with the systems shown in Figs 7.4 and 7.5. For PEAK measurement, the absolute value of the input is compared with the peak value $x_{PEAK}(n)$. If the absolute value is greater than the peak value, the difference is weighted with the coefficient AT (attack time) and added to $(1 - AT) \cdot x_{PEAK}(n - 1)$. For this attack case $|x(n)| > x_{PEAK}(n - 1)$ we get the difference equation (see Fig. 7.4)

$$x_{PEAK}(n) = (1 - AT) \cdot x_{PEAK}(n - 1) + AT \cdot |x(n)| \tag{7.11}$$

with the transfer function

$$H(z) = \frac{AT}{1 - (1 - AT)z^{-1}}. \tag{7.12}$$

If the absolute value of the input is smaller than the peak value $|x(n)| \leq x_{PEAK}(n - 1)$ (the release case), the new peak value is given by

$$x_{PEAK}(n) = (1 - RT) \cdot x_{PEAK}(n - 1) \tag{7.13}$$

with the release time coefficient RT. The difference signal of the input will be muted by the nonlinearity such that the difference equation for the peak value is given according to (7.13). For the release case the transfer function

$$H(z) = \frac{1}{1 - (1 - RT)z^{-1}} \tag{7.14}$$

is valid. For the attack case the transfer function (7.12) with coefficient AT is used and for the release case the transfer function (7.14) with the coefficient RT. The coefficients (see Section 7.3.3) are given by

$$AT = 1 - \exp\left(\frac{-2.2T_S}{t_a/1000}\right), \tag{7.15}$$

$$RT = 1 - \exp\left(\frac{-2.2T_S}{t_r/1000}\right), \tag{7.16}$$

where the attack time t_a and the release time t_r are given in milliseconds (T_S sampling interval). With this switching between filter structures one achieves fast attack responses for increasing input signal amplitudes and slow decay responses for decreasing input signal amplitudes.

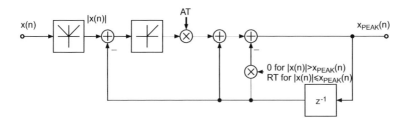

Figure 7.4 PEAK measurement.

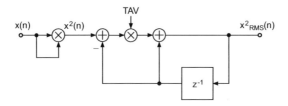

Figure 7.5 RMS measurement (TAV = averaging coefficient).

The computation of the RMS value

$$x_{\text{RMS}}(n) = \sqrt{\frac{1}{N} \sum_{i=0}^{N-1} x^2(n-i)} \qquad (7.17)$$

over N input samples can be achieved by a recursive formulation. The RMS measurement shown in Fig. 7.5 uses the square of the input and performs averaging with a first-order low-pass filter. The averaging coefficient

$$\text{TAV} = 1 - \exp\left(\frac{-2.2T_A}{t_M/1000}\right) \qquad (7.18)$$

is determined according to the time constant calculation discussed in Section 7.3.3, where t_M is the averaging time in milliseconds. The difference equation is given by

$$x_{\text{RMS}}^2(n) = (1 - \text{TAV}) \cdot x_{\text{RMS}}^2(n-1) + \text{TAV} \cdot x^2(n) \qquad (7.19)$$

with the transfer function

$$H(z) = \frac{\text{TAV}}{1 - (1 - \text{TAV})z^{-1}}. \qquad (7.20)$$

7.3.2   Gain Factor Smoothing

Attack and release times can be implemented by the system shown in Fig. 7.6 [McN84]. The attack coefficient *AT* or release coefficient *RT* is obtained by comparing the input control factor and the previous one. A small hysteresis curve determines whether the control factor is in the attack or release state and hence gives the coefficient *AT* or *RT*. The system also serves to smooth the control signal. The difference equation is given by

$$g(n) = (1 - k) \cdot g(n - 1) + k \cdot f(n),$$ (7.21)

with $k = AT$ or $k = RT$, and the corresponding transfer function leads to

$$H(z) = \frac{k}{1 - (1 - k)z^{-1}}.$$ (7.22)

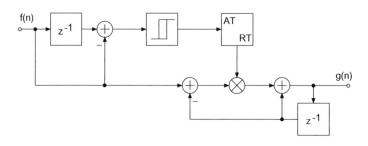

Figure 7.6 Implementing attack and release time or gain factor smoothing.

7.3.3   Time Constants

If the step response of a continuous-time system is

$$g(t) = 1 - e^{-t/\tau}, \quad \tau = \text{time constant},$$ (7.23)

then sampling (step-invariant transform) the step response gives the discrete-time step response

$$g(nT_S) = \varepsilon(nT_S) - e^{-nT_S/\tau} = 1 - z_\infty^n, \quad z_\infty = e^{-T_S/\tau}.$$ (7.24)

The Z-transform leads to

$$G(z) = \frac{z}{z - 1} - \frac{1}{1 - z_\infty z^{-1}} = \frac{1 - z_\infty}{(z - 1)(1 - z_\infty z^{-1})}.$$ (7.25)

With the definition of attack time $t_a = t_{90} - t_{10}$, we derive

$$0.1 = 1 - e^{-t_{10}/\tau} \quad \leftarrow t_{10} = 0.1\tau,$$ (7.26)
$$0.9 = 1 - e^{-t_{90}/\tau} \quad \leftarrow t_{90} = 0.9\tau.$$ (7.27)

The relationship between attack time t_a and the time constant τ of the step response is obtained as follows:

$$0.9/0.1 = e^{(t_{90}-t_{10})/\tau}$$
$$\ln(0.9/0.1) = (t_{90} - t_{10})/\tau$$
$$t_a = t_{90} - t_{10} = 2.2\tau. \qquad (7.28)$$

Hence, the pole is calculated as

$$\boxed{z_\infty = e^{-2.2T_S/t_a}.} \qquad (7.29)$$

A system for implementing the given step response is obtained by the relationship between the Z-transform of the impulse response and the Z-transform of the step response:

$$H(z) = \frac{z-1}{z}G(z). \qquad (7.30)$$

The transfer function can now be written as

$$H(z) = \frac{(1 - z_\infty)z^{-1}}{1 - z_\infty z^{-1}} \qquad (7.31)$$

with the pole $z_\infty = e^{-2.2T_S/t_a}$ adjusting the attack, release or averaging time. For the coefficients of the corresponding time constant filters the attack case is given by (7.15), the release case by (7.16) and the averaging case by (7.18). Figure 7.7 shows an example where the dotted lines mark the t_{10} and t_{90} time.

7.4 Implementation

The programming of a system for dynamic range control is described in the following sections.

7.4.1 Limiter

The block diagram of a limiter is presented in Fig. 7.8. The signal $x_{\text{PEAK}}(n)$ is determined from the input with variable attack and release time. The logarithm to the base 2 of this peak signal is taken and compared with the limiter threshold. If the signal is above the threshold, the difference is multiplied by the negative slope of the limiter LS. Then the antilogarithm of the result is taken. The control factor $f(n)$ obtained is then smoothed with a first-order low-pass filter (SMOOTH). If the signal $x_{\text{PEAK}}(n)$ lies below the limiter threshold, the signal $f(n)$ is set to $f(n) = 1$. The delayed input $x(n - D_1)$ is multiplied by the smoothed control factor $g(n)$ to give the output $y(n)$.

7.4.2 Compressor, Expander, Noise Gate

The block diagram of a compressor/expander/noise gate is shown in Fig. 7.9. The basic structure is similar to the limiter. In contrast to the limiter, the logarithm of the signal

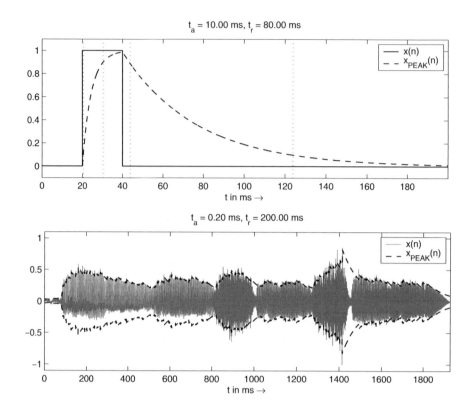

Figure 7.7 Attack and release behavior for time-constant filters.

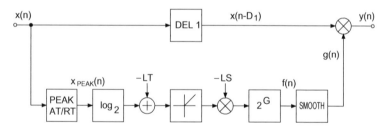

Figure 7.8 Limiter.

$x_{\mathrm{RMS}}(n)$ is taken and multiplied by 0.5. The value obtained is compared with three thresholds in order to determine the operating range of the static curve. If one of the three thresholds is crossed, the resulting difference is multiplied by the corresponding slope (CS, ES, NS) and the antilogarithm of the result is taken. A first-order low-pass filter subsequently provides the attack and release time.

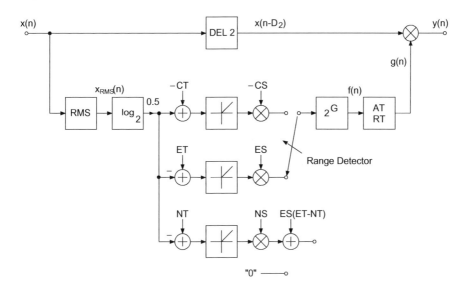

Figure 7.9 Compressor/expander/noise gate.

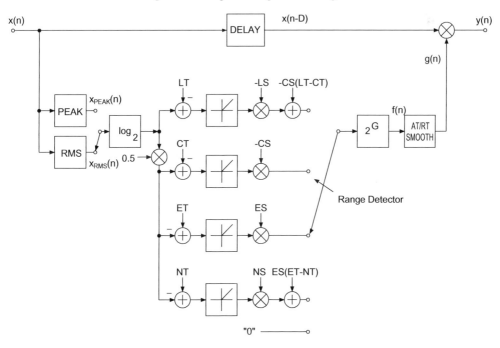

Figure 7.10 Limiter/compressor/expander/noise gate.

7.4.3   Combination System

A combination of a limiter that uses PEAK measurement, and a compressor/expander/noise gate that is based on RMS measurement, is presented in Fig. 7.10. The PEAK and RMS values are measured simultaneously. If the linear threshold of the limiter is crossed, the logarithm of the peak signal $x_{PEAK}(n)$ is taken and the upper path of the limiter is used to calculate the characteristic curve. If the limiter threshold is not crossed, the logarithm of the RMS value is taken and one of the three lower paths is used. The additive terms in the limiter and noise gate paths result from the static curve. After going through the range detector, the antilogarithm is taken. The sequence $f(n)$ is smoothed with a SMOOTH filter in the limiter case, or weighted with corresponding attack and release times of the relevant operating range (compressor, expander or noise gate). By limiting the maximum level, the dynamic range is reduced. As a consequence, the overall static curve can be shifted up by a gain factor. Figure 7.11 demonstrates this with a gain factor equal to 10 dB. This static parameter value is directly included in the control factor $g(n)$.

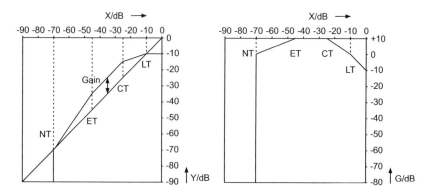

Figure 7.11 Shifting the static curve by a gain factor.

As an example, Fig. 7.12 illustrates the input $x(n)$, the output $y(n)$ and the control factor $g(n)$ of a compressor/expander system. It is observed that signals with high amplitude are compressed and those with low amplitude are expanded. An additional gain of 12 dB shows the maximum value of 4 for the control factor $g(n)$. The compressor/expander system operates in the linear region of the static curve if the control factor is equal to 4. If the control factor is between 1 and 4, the system operates as a compressor. For control factors lower than 1, the system works as an expander ($3500 < n < 4500$ and $6800 < n < 7900$). The compressor is responsible for increasing the loudness of the signal, whereas the expander increases the dynamic range for signals of small amplitude.

7.5 Realization Aspects

7.5.1 Sampling Rate Reduction

In order to reduce the computational complexity, downsampling can be carried out after calculating the PEAK/RMS value (see Fig. 7.13). As the signals $x_{PEAK}(n)$ and $x_{RMS}(n)$ are already band-limited, they can be directly downsampled by taking every second or fourth value of the sequence. This downsampled signal is then processed by taking its logarithm, calculating the static curve, taking the antilogarithm and filtering with corresponding attack

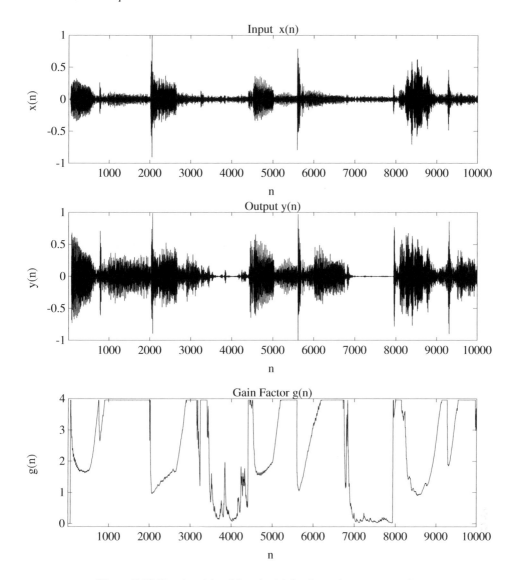

Figure 7.12 Signals $x(n)$, $y(n)$ and $g(n)$ for dynamic range control.

and release time with reduced sampling rate. The following upsampling by a factor of 4 is achieved by repeating the output value four times. This procedure is equivalent to upsampling by a factor of 4 followed by a sample-and-hold transfer function.

The nesting and spreading of partial program modules over four sampling periods is shown in Fig. 7.14. The modules PEAK/RMS (i.e. PEAK/RMS calculation) and MULT (delay of input and multiplication with $g(n)$) are performed every input sampling period. The number of processor cycles for PEAK/RMS and MULT are denoted by Z1 and Z3 respectively. The modules LD(X), CURVE, 2^x and SMO have a maximum of Z2 processor

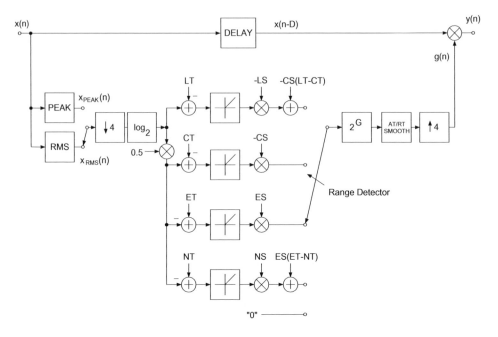

Figure 7.13 Dynamic system with sampling rate reduction.

cycles and are processed consecutively in the given order. This procedure is repeated every four sampling periods. The total number of processor cycles per sampling period for the complete dynamics algorithm results from the sum of all three modules.

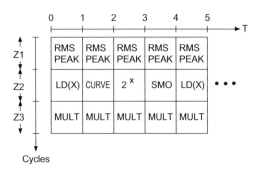

Figure 7.14 Nesting technique.

7.5.2   Curve Approximation

Besides taking logarithms and antilogarithms, other simple operations like comparisons and addition/multiplication occur in calculating the static curve. The logarithm of the

PEAK/RMS value is taken as follows:

$$x = M \cdot 2^E, \tag{7.32}$$

$$\mathrm{ld}(x) = \mathrm{ld}(M) + E. \tag{7.33}$$

First, the mantissa is normalized and the exponent is determined. The function $\mathrm{ld}(M)$ is then calculated by a series expansion. The exponent is simply added to the result.

The logarithmic weighting factor G and the antilogarithm 2^G are given by

$$G = -E - M, \tag{7.34}$$

$$2^G = 2^{-E} \cdot 2^{-M}. \tag{7.35}$$

Here, E is a natural number and M is a fractional number. The antilogarithm 2^G is calculated by expanding the function 2^{-M} in a series and multiplying by 2^{-E}. A reduction of computational complexity can be achieved by directly using log and antilog tables.

7.5.3 Stereo Processing

For stereo processing, a common control factor $g(n)$ is needed. If different control factors are used for both channels, limiting or compressing one of the two stereo signals causes a displacement of the stereo balance. Figure 7.15 shows a stereo dynamic system in which the sum of the two signals is used to calculate a common control factor $g(n)$. The following processing steps of measuring the PEAK/RMS value, downsampling, taking logarithm, calculating static curve, taking antilog attack and release time and upsampling with a sample-and-hold function remain the same. The delay (DEL) in the direct path must be the same for both channels.

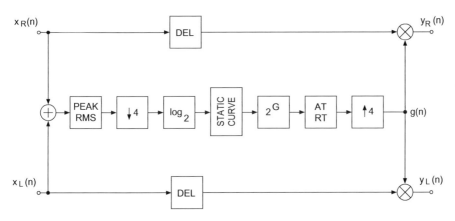

Figure 7.15 Stereo dynamic system.

7.6 Java Applet – Dynamic Range Control

The applet shown in Fig. 7.16 demonstrates dynamic range control. It is designed for a first insight into the perceptual effects of dynamic range control of an audio signal. You can adjust the characteristic curve with two control points. You can choose between two predefined audio files from our web server (*audio1.wav* or *audio2.wav*) or your own local wav file to be processed [Gui05].

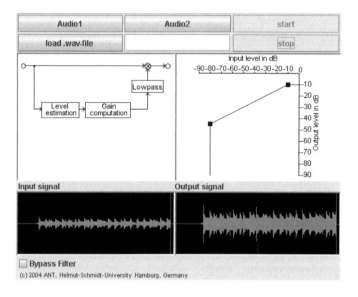

Figure 7.16 Java applet – dynamic range control.

7.7 Exercises

1. Low-pass Filtering for Envelope Detection

Generally, envelope computation is performed by low-pass filtering the input signal's absolute value or its square.

1. Sketch the block diagram of a recursive first-order low-pass $H(z) = \lambda/[(1 - (1 - \lambda)z^{-1})]$.

2. Sketch its step response. What characteristic measure of the envelope detector can be derived from the step response and how?

3. Typically, the low-pass filter is modified to use a non-constant filter coefficient λ. How does λ depend on the signal? Sketch the response to a rect signal of the low-pass filter thus modified.

2. Discrete-time Specialties of Envelope Detection

Taking absolute value or squaring are non-linear operations. Therefore, care must be taken when using them in discrete-time systems as they introduce harmonics the frequency of which may violate the Nyquist bound. This can lead to unexpected results, as a simple example illustrates. Consider the input signal $x(n) = \sin(\frac{\pi}{2}n + \varphi)$, $\varphi \in [0, 2\pi]$.

1. Sketch $x(n)$, $|x(n)|$ and $x^2(n)$ for different values of φ.

2. Determine the value of the envelope after perfect low-pass filtering, i.e. averaging, $|x(n)|$. Note: As the input signal is periodical, it is sufficient to consider one period, e.g.

$$\bar{x} = \frac{1}{4} \sum_{n=0}^{3} |x(n)|.$$

3. Similarly, determine the value of the envelope after averaging $x^2(n)$.

3. Dynamic Range Processors

Sketch the characteristic curves mapping input level to output level and input level to gain for and describe briefly the application of:

1. limiter;

2. compressor;

3. expander;

4. noise gate.

References

[Gui05]   M. Guillemard, C. Ruwwe, U. Zölzer: *J-DAFx – Digital Audio Effects in Java*, Proc. 8th Int. Conference on Digital Audio Effects (DAFx-05), pp. 161–166, Madrid, 2005.

[McN84]   G. W. McNally: *Dynamic Range Control of Digital Audio Signals*, J. Audio Eng. Soc., Vol. 32, pp. 316–327, 1984.

[Sti86]   E. Stikvoort: *Digital Dynamic Range Compressor for Audio*, J. Audio Eng. Soc., Vol. 34, pp. 3–9, 1986.

Chapter 8

Sampling Rate Conversion

Several different sampling rates are established for digital audio applications. For broadcasting, professional and consumer audio, sampling rates of 32, 48 and 44.1 kHz are used. Moreover, other sampling rates are derived from different frame rates for film and video. In connecting systems with different uncoupled sampling rates, there is a need for sampling rate conversion. In this chapter, synchronous sampling rate conversion with rational factor L/M for coupled clock rates and asynchronous sampling rate conversion will be discussed where the different sampling rates are not synchronized with each other.

8.1 Basics

Sampling rate conversion consists out of upsampling and downsampling and anti-imaging and anti-aliasing filtering [Cro83, Vai93, Fli00, Opp99]. The discrete-time Fourier transform of the sampled signal $x(n)$ with sampling frequency $f_S = 1/T$ ($\omega_S = 2\pi f_S$) is given by

$$X(e^{j\Omega}) = \frac{1}{T} \sum_{k=-\infty}^{\infty} X_a\left(j\omega + jk \underbrace{\frac{2\pi}{T}}_{\omega_S} \right), \quad \Omega = \omega T, \tag{8.1}$$

with the Fourier transform $X_a(j\omega)$ of the continuous-time signal $x(t)$. For ideal sampling the condition

$$X(e^{j\Omega}) = \frac{1}{T} X_a(j\omega), \quad |\Omega| \leq \pi, \tag{8.2}$$

holds.

8.1.1 Upsampling and Anti-imaging Filtering

For upsampling the signal

$$x(n) \circ\!\!\!-\!\!\!\bullet X(e^{j\Omega}) \tag{8.3}$$

Digital Audio Signal Processing Second Edition   Udo Zölzer
© 2008 John Wiley & Sons, Ltd

by a factor L between consecutive samples $L - 1$ zero samples will be included (see Fig. 8.1). This leads to the upsampled signal

$$w(m) = \begin{cases} x\left(\dfrac{m}{L}\right), & m = 0, \pm L, \pm 2L, \ldots, \\ 0, & \text{otherwise}, \end{cases} \tag{8.4}$$

with sampling frequency $f_S' = 1/T' = L \cdot f_S = L/T$ $(\Omega' = \Omega/L)$ and the corresponding Fourier transform

$$W(e^{j\Omega'}) = \sum_{m=-\infty}^{\infty} w(m)\, e^{-jm\Omega'} = \sum_{m=-\infty}^{\infty} x(m)\, e^{-jmL\Omega'} = X(e^{jL\Omega'}). \tag{8.5}$$

The suppression of the image spectra is achieved by anti-imaging filtering of $w(m)$ with $h(m)$, such that the output signal is given by

$$y(m) = w(m) * h(m), \tag{8.6}$$

$$Y(e^{j\Omega'}) = H(e^{j\Omega'}) \cdot X(e^{j\Omega'L}). \tag{8.7}$$

To adjust the signal power in the base-band the Fourier transform of the impulse-response

$$H(e^{j\Omega'}) = \begin{cases} L, & |\Omega'| \le \pi/L, \\ 0, & \text{otherwise}, \end{cases} \tag{8.8}$$

needs a gain factor L in the pass-band, such that the output signal $y(m)$ has the Fourier transform given by

$$Y(e^{j\Omega'}) = LX(e^{j\Omega'L}) \tag{8.9}$$

$$= L\,\underbrace{\frac{1}{T} \sum_{k=-\infty}^{\infty} X_a\left(j\omega + jLk\frac{2\pi}{T}\right)}_{\text{with (8.1) and (8.5)}} \tag{8.10}$$

$$= L\,\frac{1}{LT'} \sum_{k=-\infty}^{\infty} X_a\left(j\omega + jLk\frac{2\pi}{LT'}\right) \tag{8.11}$$

$$= \underbrace{\frac{1}{T'} \sum_{k=-\infty}^{\infty} X_a\left(j\omega + jk\frac{2\pi}{T'}\right)}_{\text{spectrum of signal with } f_S'=Lf_S}. \tag{8.12}$$

The output signal represents the sampling of the input $x(t)$ with sampling frequency $f_S' = Lf_S$.

8.1.2 Downsampling and Anti-aliasing Filtering

For downsampling a signal $x(n)$ by M the signal has to be band-limited to π/M in order to avoid aliasing after the downsampling operation (see Fig. 8.2). Band-limiting is achieved

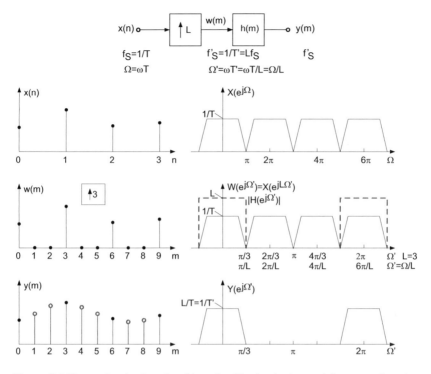

Figure 8.1 Upsampling by L and anti-imaging filtering in time and frequency domain.

by filtering with $H(e^{j\Omega})$ according to

$$w(m) = x(m) * h(m), \tag{8.13}$$

$$W(e^{j\Omega}) = X(e^{j\Omega}) \cdot H(e^{j\Omega}), \tag{8.14}$$

$$H(e^{j\Omega}) = \begin{cases} 1, & |\Omega| \leq \pi/M, \\ 0, & \text{otherwise.} \end{cases} \tag{8.15}$$

Downsampling of $w(m)$ is performed by taking every Mth sample, which leads to the output signal

$$y(n) = w(Mn) \tag{8.16}$$

with the Fourier transform

$$Y(e^{j\Omega'}) = \frac{1}{M} \sum_{l=0}^{M-1} W(e^{j(\Omega'-2\pi l)/M}). \tag{8.17}$$

For the base-band spectrum ($|\Omega'| \leq \pi$ and $l = 0$) we get

$$Y(e^{j\Omega'}) = \frac{1}{M} H(e^{j\Omega'/M}) \cdot X(e^{j\Omega'/M}) = \frac{1}{M} X(e^{j\Omega'/M}), \quad |\Omega'| \leq \pi, \tag{8.18}$$

and for the Fourier transform of the output signal we can derive

$$Y(e^{j\Omega'}) = \frac{1}{M} X(e^{j\Omega'/M}) = \underbrace{\frac{1}{M} \frac{1}{T} \sum_{k=-\infty}^{\infty} X_a\left(j\omega + jk\frac{2\pi}{MT}\right)}_{\text{with (8.1)}} \tag{8.19}$$

$$= \underbrace{\frac{1}{T'} \sum_{k=-\infty}^{\infty} X_a\left(j\omega + jk\frac{2\pi}{T'}\right)}_{\text{spectrum of signal with } f'_S = f_S/M}, \tag{8.20}$$

which represents a sampled signal $y(n)$ with $f'_S = f_S/M$.

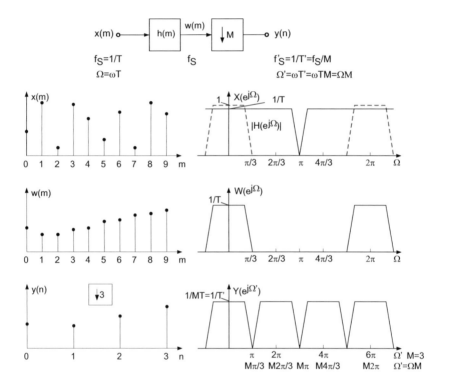

Figure 8.2 Anti-aliasing filtering and downsampling by M in time and frequency domain.

8.2   Synchronous Conversion

Sampling rate conversion for coupled sampling rates by a rational factor L/M can be performed by the system shown in Fig. 8.3. After upsampling by a factor L, anti-imaging filtering at Lf_S is carried out, followed by downsampling by factor M. Since after upsampling and filtering only every Mth sample is used, it is possible to develop efficient algorithms that reduce complexity. In this respect two methods are in use: one is based on a time-domain interpretation [Cro83] and the other [Hsi87] uses Z-domain fundamentals. Owing to its computational efficiency, only the method in the Z-domain will be considered.

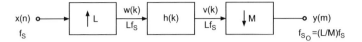

Figure 8.3 Sampling rate conversion by factor L/M.

Starting with the finite impulse response $h(n)$ of length N and its Z-transform

$$H(z) = \sum_{n=0}^{N-1} h(n)z^{-n},\tag{8.21}$$

the polyphase representation [Cro83, Vai93, Fli00] with M components can be expressed as

$$H(z) = \sum_{k=0}^{M-1} z^{-k} E_k(z^M)\tag{8.22}$$

with

$$e_k(n) = h(nM + k), \quad k = 0, 1, \ldots, M-1,\tag{8.23}$$

or

$$H(z) = \sum_{k=0}^{M-1} z^{-(M-1-k)} R_k(z^M)\tag{8.24}$$

with

$$r_k(n) = h(nM - k), \quad k = 0, 1, \ldots, M-1.\tag{8.25}$$

The polyphase decomposition as given in (8.22) and (8.24) is referred to as type 1 and 2, respectively. The type 1 polyphase decomposition corresponds to a commutator model in the anti-clockwise direction whereas the type 2 is in the clockwise direction. The relationship between $R(z)$ and $E(z)$ is described by

$$R_k(z) = E_{M-1-k}(z).\tag{8.26}$$

With the help of the identities [Vai93] shown in Fig. 8.4 and the decomposition (Euclid's theorem)

$$z^{-1} = z^{-pL} z^{qM},\tag{8.27}$$

it is possible to move the inner delay elements of Fig. 8.5. Equation (8.27) is valid if M and L are prime numbers. In a cascade of upsampling and downsampling, the order of functional blocks can be exchanged (see Fig. 8.5b).

The use of polyphase decomposition can be demonstrated with the help of an example for $L = 2$ and $M = 3$. This implies a sampling rate conversion from 48 kHz to 32 kHz. Figures 8.6 and 8.7 show two different solutions for polyphase decomposition of sampling rate conversion by 2/3. Further decompositions of the upsampling decomposition of Fig. 8.7 are demonstrated in Fig. 8.8. First, interpolation is implemented with a polyphase

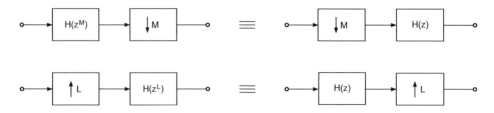

Figure 8.4 Identities for sampling rate conversion.

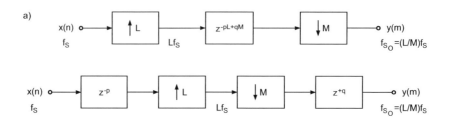

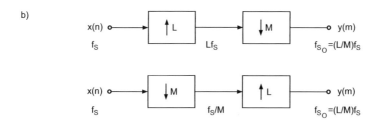

Figure 8.5 Decomposition in accordance with Euclid's theorem.

decomposition and the delay z^{-1} is decomposed to $z^{-1} = z^{-2}z^3$. Then, the downsampler of factor 3 is moved through the adder into the two paths (Fig. 8.8b) and the delays are moved according to the identities of Fig. 8.4. In Fig. 8.8c, the upsampler is exchanged with the downsampler, and in a final step (Fig. 8.8d) another polyphase decomposition of $E_0(z)$ and $E_1(z)$ is carried out. The actual filter operations $E_{0k}(z)$ and $E_{1k}(z)$ with $k = 0, 1, 2$ are performed at $\frac{1}{3}$ of the input sampling rate.

8.3 Asynchronous Conversion

Plesiochronous systems consist of partial systems with different and uncoupled sampling rates. Sampling rate conversion between such systems can be achieved through a DA conversion with the sampling rate of the first system followed by an AD conversion with the sampling rate of the second system. A digital approximation of this approach is made with a multirate system [Lag81, Lag82a, Lag82b, Lag82c, Lag83, Ram82, Ram84]. Figure 9.9a shows a system for increasing the sampling rate by a factor L followed by an anti-imaging

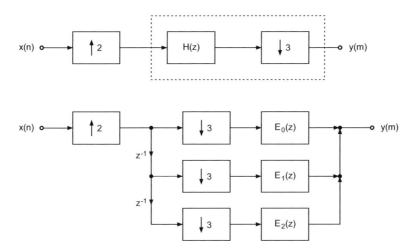

Figure 8.6 Polyphase decomposition for downsampling $L/M = 2/3$.

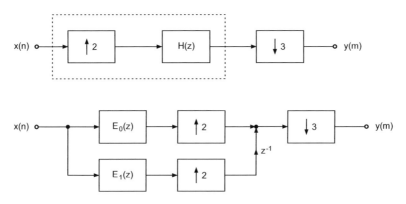

Figure 8.7 Polyphase decomposition for upsampling $L/M = 2/3$.

filter $H(z)$ and a resampling of the interpolated signal $y(k)$. The samples $y(k)$ are held for a clock period (see Fig. 8.9c) and then sampled with output clock period $T_{S_O} = 1/f_{S_O}$. The interpolation sampling rate must be increased sufficiently that the difference of two consecutive samples $y(k)$ is smaller than the quantization step Q. The sample-and-hold function applied to $y(k)$ suppresses the spectral images at multiples of Lf_S (see Fig. 8.9b). The signal obtained is a band-limited continuous-time signal which can be sampled with output sampling rate f_{S_O}.

For the calculation of the necessary oversampling rate, the problem is considered in the frequency domain. The sinc function of a sample-and-hold system (see Fig. 8.9b) at frequency $\tilde{f} = (L - \frac{1}{2})f_S$ is given by

$$E(\tilde{f}) = \frac{\sin\left(\frac{\pi \tilde{f}}{Lf_S}\right)}{\frac{\pi \tilde{f}}{Lf_S}} = \frac{\sin\left(\frac{\pi(L-\frac{1}{2})f_S}{Lf_S}\right)}{\frac{\pi(L-\frac{1}{2})f_S}{Lf_S}} = \frac{\sin\left(\pi - \frac{\pi}{2L}\right)}{\pi - \frac{\pi}{2L}}. \tag{8.28}$$

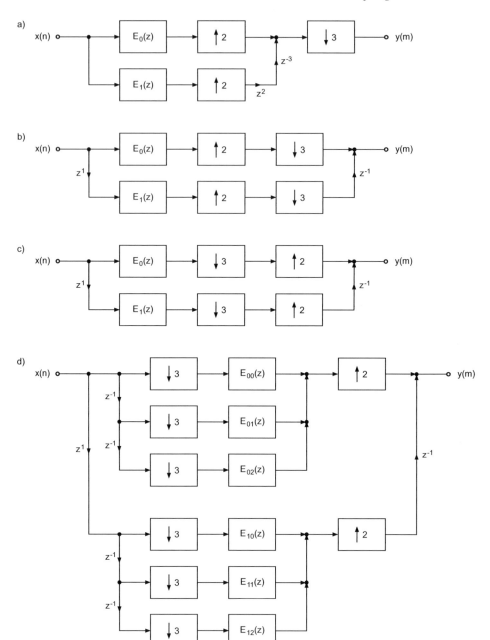

Figure 8.8 Sampling rate conversion by factor 2/3.

With $\sin(\alpha - \beta) = \sin(\alpha)\cos(\beta) - \cos(\alpha)\sin(\beta)$ we derive

$$E(\tilde{f}) = \frac{\sin\left(\frac{\pi}{2L}\right)}{\pi\left(1 - \frac{1}{2L}\right)} \approx \frac{\pi/2L}{\pi\left(1 - \frac{1}{2L}\right)} \approx \frac{1}{2L - 1} \approx \frac{1}{2L}. \tag{8.29}$$

a)
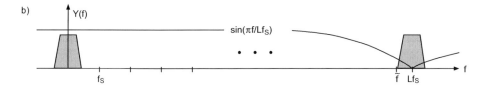

b)
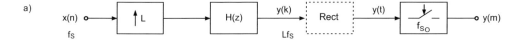

c)
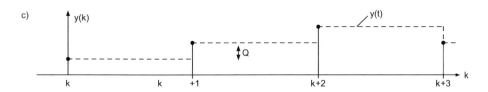

Figure 8.9 Approximation of DA/AD conversions.

This value of (8.29) should be lower than $\frac{Q}{2}$ and allows the computation of the interpolation factor L. For a given word-length w and quantization step Q, the necessary interpolation rate L is calculated by

$$\frac{Q}{2} \geq \frac{1}{2L}, \tag{8.30}$$

$$\frac{2^{-(w-1)}}{2} \geq \frac{1}{2L} \tag{8.31}$$

$$\hookrightarrow L \geq 2^{w-1}. \tag{8.32}$$

For a linear interpolation between upsampled samples $y(k)$, we can derive

$$E(\tilde{f}) = \frac{\sin^2\left(\frac{\pi\tilde{f}}{Lf_S}\right)}{\left(\frac{\pi\tilde{f}}{Lf_S}\right)^2} \tag{8.33}$$

$$= \frac{\sin^2\left(\frac{\pi(L-\frac{1}{2})f_S}{Lf_S}\right)}{\left(\frac{\pi(L-\frac{1}{2})f_S}{Lf_S}\right)^2} \tag{8.34}$$

$$\approx \frac{1}{(2L)^2}. \tag{8.35}$$

With this it is possible to reduce the necessary interpolation rate to

$$L_1 \geq 2^{w/2-1}. \tag{8.36}$$

Figure 8.10 demonstrates this with a two-stage block diagram. First, interpolation up to a sampling rate $L_1 f_S$ is performed by conventional filtering. In a second stage upsampling by factor L_2 is done by linear interpolation. The two-stage approach must satisfy the sampling rate $L f_S = (L_1 L_2) f_S$.

The choice of the interpolation algorithm in the second stage enables the reduction of the first oversampling factor. More details are discussed in Section 8.2.2.

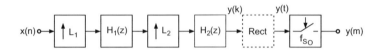

Figure 8.10 Linear interpolation before virtual sample-and-hold function.

8.3.1   Single-stage Methods

Direct conversion methods implement the block diagram [Lag83, Smi84, Par90, Par91a, Par91b, Ada92, Ada93] shown in Fig. 8.9a. The calculation of a discrete sample on an output grid of sampling rate f_{S_O} from samples $x(n)$ at sampling rate f_{S_I} can be written as

$$\mathrm{DFT}[x(n-\alpha)] = X(e^{j\Omega})\, e^{-j\alpha\Omega} = X(e^{j\Omega}) H_\alpha(e^{j\Omega}), \tag{8.37}$$

where $0 \leq \alpha < 1$. With the transfer function

$$H_\alpha(e^{j\Omega}) = e^{-j\alpha\Omega} \tag{8.38}$$

and the properties

$$H(e^{j\Omega}) = \begin{cases} 1, & 0 \leq |\Omega| \leq \Omega_c, \\ 0, & \Omega_c < |\Omega| < \pi, \end{cases} \tag{8.39}$$

the impulse response is given by

$$h_\alpha(n) = h(n-\alpha) = \frac{\Omega_c}{\pi}\frac{\sin[\Omega_c(n-\alpha)]}{\Omega_c(n-\alpha)}. \tag{8.40}$$

From (8.37) we can express the delayed signal

$$x(n-\alpha) = \sum_{m=-\infty}^{\infty} x(m) h(n-\alpha-m) \tag{8.41}$$

$$= \sum_{m=-\infty}^{\infty} x(m) \frac{\Omega_c}{\pi}\frac{\sin[\Omega_c(n-\alpha-m)]}{\Omega_c(n-\alpha-m)} \tag{8.42}$$

as the convolution between $x(n)$ and $h(n-\alpha)$. Figure 8.11 illustrates this convolution in the time domain for a fixed α. Figure 8.12 shows the coefficients $h(n-\alpha_i)$ for discrete

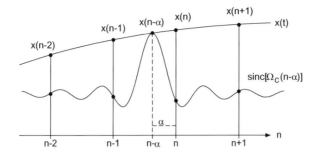

Figure 8.11 Convolution sum (8.42) in the time domain.

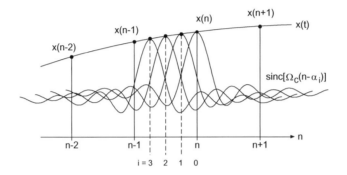

Figure 8.12 Convolution sum (8.42) for different α_i.

α_i ($i = 0, \ldots, 3$) which are obtained from the intersection of the sinc function with the discrete samples $x(n)$.

In order to limit the convolution sum, the impulse response is windowed, which gives

$$h_W(n - \alpha_i) = w(n)\frac{\Omega_c}{\pi}\frac{\sin[\Omega_c(n - \alpha_i)]}{\Omega_c(n - \alpha_i)}, \quad n = 0, \ldots, 2M. \tag{8.43}$$

From this, the sample estimate

$$\hat{x}(n - \alpha_i) = \sum_{m=-M}^{M} x(m)h_W(n - \alpha_i - m) \tag{8.44}$$

results. A graphical interpretation of the time-variant impulse response which depends on α_i is shown in Fig. 8.13. The discrete segmentation between two input samples into N intervals leads to N partial impulse responses of length $2M + 1$.

If the output sampling rate is smaller than the input sampling rate ($f_{S_O} < f_{S_I}$), band-limiting (anti-aliasing) to the output sampling rate has to be done. This can be achieved with factor $\beta = f_{S_O}/f_{S_I}$ and leads, with the scaling theorem of the Fourier transform, to

$$h(n - \alpha) = \frac{\beta\Omega_c}{\pi}\frac{\sin[\beta\Omega_c(n - \alpha)]}{\beta\Omega_c(n - \alpha)}. \tag{8.45}$$

This time-scaling of the impulse response has the consequence that the number of coefficients of the time-variant partial impulse responses is increased. The number of required states also increases. Figure 8.14 shows the time-scaled impulse response and elucidates the increase in the number M of the coefficients.

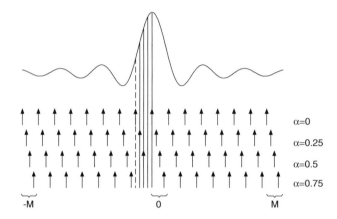

Figure 8.13 Sinc function and different impulse responses.

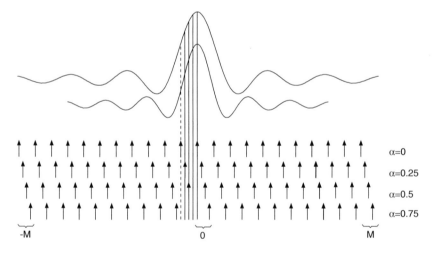

Figure 8.14 Time-scaled impulse response.

8.3.2 Multistage Methods

The basis of a multistage conversion method [Lag81, Lag82, Kat85, Kat86] is shown in Fig. 8.15a and will be described in the frequency domain as shown in Fig. 8.15b–d. Increasing the sampling rate up to Lf_S before the sample-and-hold function is done in four stages. In the first two stages, the sampling rate is increased by a factor of 2 followed by an anti-imaging filter (see Fig. 8.15b,c), which leads to a four times oversampled spectrum

(Fig. 8.15d). In the third stage, the signal is upsampled by a factor of 32 and the image spectra are suppressed (see Fig. 8.15d,e). In the fourth stage (Fig. 8.15e) the signal is upsampled to a sampling rate of Lf_S by a factor of 256 and a linear interpolator. The sinc^2 function of the linear interpolator suppresses the images at multiples of $128 f_S$ up to the spectrum at Lf_S. The virtual sample-and-hold function is shown in Fig. 8.15f, where resampling at the output sampling rate is performed. A direct conversion of this kind of cascaded interpolation structure requires anti-imaging filtering after every upsampling with the corresponding sampling rate. Although the necessary filter order decreases owing to a decrease in requirements for filter design, an implementation of the filters in the third and fourth stages is not possible directly. Following a suggestion by Lagadec [Lag82c], the measurement of the ratio of input to output rate is used to control the polyphase filters in the third and fourth stages (see Fig. 8.16a, CON = control) to reduce complexity. Figures 8.16b–d illustrate an interpretation in the time domain. Figure 8.16b shows the interpolation of three samples between two input samples $x(n)$ with the help of the first and second interpolation stage. The abscissa represents the intervals of the input sampling rate and the sampling rate is increased by factor of 4. In Fig. 8.16c the four times oversampled signal is shown. The abscissa shows the four times oversampled output grid. It is assumed that output sample $y(m = 0)$ and input sample $x(n = 0)$ are identical. The output sample $y(m = 1)$ is now determined in such a form that with the interpolator in the third stage only two polyphase filters just before and after the output sample need to be calculated. Hence, only two out of a total of 31 possible polyphase filters are calculated in the third stage. Figure 8.16d shows these two polyphase output samples. Between these two samples, the output sample $y(m = 1)$ is obtained with a linear interpolation on a grid of 255 values.

Instead of the third and fourth stages, special interpolation methods can be used to calculate the output $y(m)$ directly from the four times oversampled input signal (see Fig. 8.17) [Sti91, Cuc91, Liu92]. The upsampling factor $L_3 = 2^{w-3}$ for the last stage is calculated according to $L = 2^{w-1} = L_1 L_2 L_3 = 2^2 L_3$. Section 8.4 is devoted to different interpolation methods which allow a real-time calculation of filter coefficients. This can be interpreted as time-variant filters in which the filter coefficients are derived from the ratio of sampling rates. The calculation of one filter coefficient set for the output sample at the output rate is done by measuring the ratio of input to output sampling rate as described in the next section.

8.3.3 Control of Interpolation Filters

The measurement of the ratio of input and output sampling rate is used for controlling the interpolation filters [Lag82a]. By increasing the sampling rate by a factor of L the input sampling period is divided into $L = 2^{w-1} = 2^{15}$ parts for a signal word-length of $w = 16$ bits. The time instant of the output sample is calculated on this grid with the help of the measured ratio of sampling periods T_{S_O}/T_{S_I} as follows.

A counter is clocked with Lf_{S_I} and reset by every new input sampling clock. A sawtooth curve of the counter output versus time is obtained as shown in Fig. 8.18. The counter runs from 0 to $L - 1$ during one input sampling period. The output sampling period T_{S_O} starts at time t_{i-2}, which corresponds to counter output z_{i-2}, and stops at time t_{i-1}, with counter output z_{i-1}. The difference between both counter measurements allows the calculation of the output sampling period T_{S_O} with a resolution of Lf_{S_I}.

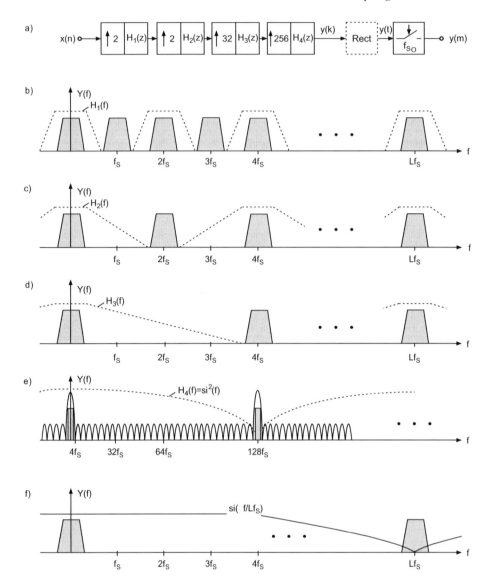

Figure 8.15 Multistage conversion – frequency-domain interpretation.

The new counter measurement is added to the difference of previous counter measurements. As a result, the new counter measurement is obtained as

$$t_i = (t_{i-1} + T_{S_O}) \oplus T_{S_I}. \tag{8.46}$$

The modulo operation can be carried out with an accumulator of word-length $w - 1 = 15$. The resulting time t_i determines the time instant of the output sample at the output sampling rate and therefore the choice of the polyphase filter in a single-stage conversion or the time instant for a multistage conversion.

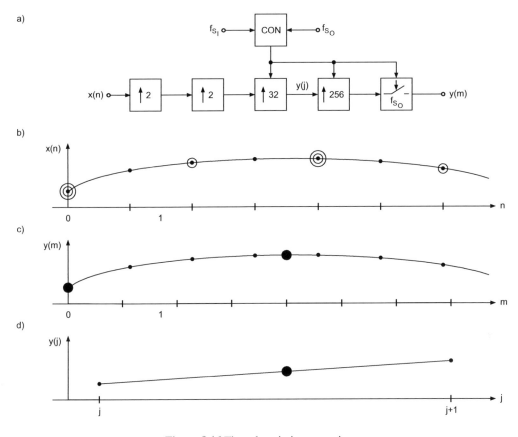

Figure 8.16 Time-domain interpretation.

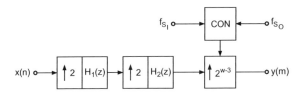

Figure 8.17 Sampling rate conversion with interpolation for calculating coefficients of a time-variant interpolation filter.

The measurement of T_{S_O}/T_{S_I} is illustrated in Fig. 8.19:

- The input sampling rate f_{S_I} is increased to $M_Z f_{S_I}$ using a frequency multiplier where $M_Z = 2^w$. This input clock increase by the factor M_Z triggers a w-bit counter. The counter output z is evaluated every M_O output sampling periods.

- Counting of M_O output sampling periods.

- Simultaneous counting of the M_I input sampling periods.

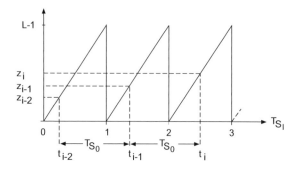

Figure 8.18 Calculation of t_i.

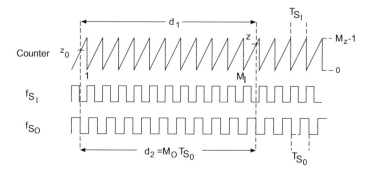

Figure 8.19 Measurement of T_{S_O}/T_{S_I}.

The time intervals d_1 and d_2 (see Fig. 8.19) are given by

$$d_1 = M_I T_{S_I} + \frac{z - z_0}{M_Z} T_{S_I} = \left(M_I + \frac{z - z_0}{M_Z} \right) T_{S_I}, \tag{8.47}$$

$$d_2 = M_O T_{S_O}, \tag{8.48}$$

and with the requirement $d_1 = d_2$ we can write

$$M_O T_{S_O} = \left(M_I + \frac{z - z_0}{M_Z} \right) T_{S_I}$$

$$\frac{T_{S_O}}{T_{S_I}} = \frac{M_I + (z - z_0)/M_Z}{M_O} = \frac{M_Z M_I + (z - z_0)}{M_Z M_O}. \tag{8.49}$$

- Example 1: $w = 0 \rightarrow M_Z = 1$

$$\frac{T_{S_O}}{T_{S_I}} = \frac{M_I}{2^{15}} \tag{8.50}$$

With a precision of 15 bits, the averaging number is chosen as $M_O = 2^{15}$ and the number M_I has to be determined.

- Example 2: $w = 8 \rightarrow M_Z = 2^8$

$$\frac{T_{S_O}}{T_{S_I}} = \frac{2^8 M_I + (z - z_0)}{2^8 2^7} \tag{8.51}$$

With a precision of 15 bits, the averaging number is chosen as $M_O = 2^7$ and the number M_I and the counter outputs have to be determined.

The sampling rates at the input and output of a sampling rate converter can be calculated by evaluating the 8-bit increment of the counter for each output clock with

$$z = \frac{T_{S_O}}{T_{S_I}} M_Z = \frac{f_{S_I}}{f_{S_O}} 256, \tag{8.52}$$

as seen from Table 8.1.

Table 8.1 Counter increments for different sampling rate conversions.

Conversion/kHz	8-bit counter increment
32 → 48	170
44.1 → 48	235
32 → 44.1	185
48 → 44.1	278
48 → 32	384
44.1 → 32	352

8.4   Interpolation Methods

In the following sections, special interpolation methods are discussed. These methods enable the calculation of time-variant filter coefficients for sampling rate conversion and need an oversampled input sequence as well as the time instant of the output sample. A convolution of the oversampled input sequence with time-variant filter coefficients gives the output sample at the output sampling rate. This real-time computation of filter coefficients is not based on popular filter design methods. On the contrary, methods are presented for calculating filter coefficient sets for every input clock cycle where the filter coefficients are derived from the distance of output samples to the time grid of the oversampled input sequence.

8.4.1   Polynomial Interpolation

The aim of a polynomial interpolation [Liu92] is to determine a polynomial

$$p_N(x) = \sum_{i=0}^{N} a_i x^i \tag{8.53}$$

of Nth order representing exactly a function $f(x)$ at $N + 1$ uniformly spaced x_i, i.e. $p_N(x_i) = f(x_i) = y_i$ for $i = 0, \ldots, N$. This can be written as a set of linear equations

$$
\begin{bmatrix}
1 & x_0 & x_0^2 & \cdots & x_0^N \\
1 & x_1 & x_1^2 & \cdots & x_1^N \\
\vdots & \vdots & \vdots & & \vdots \\
1 & x_N & x_N^N & \cdots & x_N^N
\end{bmatrix}
\begin{bmatrix}
a_0 \\
a_1 \\
\vdots \\
a_N
\end{bmatrix}
=
\begin{bmatrix}
y_0 \\
y_1 \\
\vdots \\
y_N
\end{bmatrix}.
\tag{8.54}
$$

The polynomial coefficients a_i as functions of $y_0, \ldots, y_N$ are obtained with the help of Cramer's rule according to

$$
a_i = \frac{
\begin{vmatrix}
1 & x_0 & x_0^2 & \cdots & y_0 & \cdots & x_0^N \\
1 & x_1 & x_1^2 & \cdots & y_1 & \cdots & x_1^N \\
\vdots & \vdots & \vdots & & \vdots & \cdots & \vdots \\
1 & x_N & x_N^2 & \cdots & y_N & \cdots & x_N^N
\end{vmatrix}
}{
\begin{vmatrix}
1 & x_0 & x_0^2 & \cdots & x_0^N \\
1 & x_1 & x_1^2 & \cdots & x_1^N \\
\vdots & \vdots & \vdots & & \vdots \\
1 & x_N & x_N^2 & \cdots & x_N^N
\end{vmatrix}
}, \qquad i = 0, 1, \ldots, N.
\tag{8.55}
$$

For uniformly spaced $x_i = i$ with $i = 0, 1, \ldots, N$ the interpolation of an output sample with distance α gives

$$
y(n + \alpha) = \sum_{i=0}^{N} a_i (n + \alpha)^i.
\tag{8.56}
$$

In order to determine the relationship between the output sample $y(n + \alpha)$ and y_i, a set of time-variant coefficients c_i needs to be determined such that

$$
\boxed{y(n + \alpha) = \sum_{i=-N/2}^{N/2} c_i(\alpha) y(n + i).}
\tag{8.57}
$$

The calculation of time-variant coefficients $c_i(\alpha)$ will be illustrated by an example.

Example: Figure 8.20 shows the interpolation of an output sample of distance α with $N = 2$ and using three samples which can be written as

$$
y(n + \alpha) = \sum_{i=0}^{2} a_i (n + \alpha)^i.
\tag{8.58}
$$

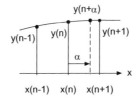

Figure 8.20 Polynomial interpolation with three samples.

The samples $y(n + i)$, with $i = -1, 0, 1$, can be expressed as

$$y(n + 1) = \sum_{i=0}^{2} a_i (n + 1)^i, \quad \alpha = 1,$$

$$y(n) = \sum_{i=0}^{2} a_i n^i, \quad \alpha = 0,$$

$$y(n - 1) = \sum_{i=0}^{2} a_i (n - 1)^i, \quad \alpha = -1, \tag{8.59}$$

or in matrix notation

$$\begin{bmatrix} 1 & (n+1) & (n+1)^2 \\ 1 & n & n^2 \\ 1 & (n-1) & (n-1)^2 \end{bmatrix} \begin{bmatrix} a_0 \\ a_1 \\ a_2 \end{bmatrix} = \begin{bmatrix} y(n+1) \\ y(n) \\ y(n-1) \end{bmatrix}. \tag{8.60}$$

The coefficients a_i as functions of y_i are then given by

$$\begin{bmatrix} a_0 \\ a_1 \\ a_2 \end{bmatrix} = \begin{bmatrix} \dfrac{n(n-1)}{2} & 1 - n^2 & \dfrac{n(n+1)}{2} \\ -\dfrac{2n-1}{2} & 2n & -\dfrac{2n+1}{2} \\ \dfrac{1}{2} & -1 & \dfrac{1}{2} \end{bmatrix} \begin{bmatrix} y(n+1) \\ y(n) \\ y(n-1) \end{bmatrix}, \tag{8.61}$$

such that

$$y(n + \alpha) = a_0 + a_1 (n + \alpha) + a_2 (n + \alpha)^2 \tag{8.62}$$

is valid. The output sample $y(n + \alpha)$ can be written as

$$y(n + \alpha) = \sum_{i=-1}^{1} c_i(\alpha) y(n + i)$$

$$= c_{-1} y(n - 1) + c_0 y(n) + c_1 y(n + 1). \tag{8.63}$$

Equation (8.62) with a_i from (8.61) leads to

$$
y(n + \alpha) = \left[\frac{1}{2} y(n + 1) - y(n) + \frac{1}{2} y(n - 1) \right](n + \alpha)^2
$$

$$
+ \left[-\frac{2n - 1}{2} y(n + 1) + 2ny(n) - \frac{2n + 1}{2} y(n - 1) \right](n + \alpha)
$$

$$
+ \frac{n(n - 1)}{2} y(n + 1) + (1 - n^2)y(n) + \frac{n(n + 1)}{2} y(n - 1). \qquad (8.64)
$$

Comparing the coefficients from (8.63) and (8.64) for $n = 0$ gives the coefficients

$$
c_{-1} = \tfrac{1}{2}\alpha(\alpha - 1),
$$

$$
c_0 = -(\alpha - 1)(\alpha + 1) = 1 - \alpha^2,
$$

$$
c_1 = \tfrac{1}{2}\alpha(\alpha + 1).
$$

8.4.2 Lagrange Interpolation

Lagrange interpolation for $N + 1$ samples makes use of the polynomials $l_i(x)$ which have the following properties (see Fig. 8.21):

$$
l_i(x_k) = \delta_{ik} = \begin{cases} 1, & i = k, \\ 0, & \text{elsewhere.} \end{cases} \qquad (8.65)
$$

Based on the zeros of the polynomial $l_i(x)$, it follows that

$$
l_i(x) = a_i(x - x_0) \cdots (x - x_{i-1})(x - x_{i+1}) \cdots (x - x_N). \qquad (8.66)
$$

With $l_i(x_i) = 1$ the coefficients are given by

$$
a_i(x_i) = \frac{1}{(x_i - x_0) \cdots (x_i - x_{i-1})(x_i - x_{i+1}) \cdots (x_i - x_N)}. \qquad (8.67)
$$

The interpolation polynomial is expressed as

$$
p_N(x) = \sum_{i=0}^{N} l_i(x) y_i = l_0(x)y_0 + \cdots + l_N(x)y_N. \qquad (8.68)
$$

With $a = \prod_{j=0}^{N}(x - x_j)$, (8.66) can be written as

$$
l_i(x) = a_i \frac{a}{x - x_i} = \frac{1}{\prod_{j=0, j \neq i}^{N} x_i - x_j} \frac{\prod_{j=0}^{N} x - x_j}{x - x_i} = \prod_{j=0, j \neq i}^{N} \frac{x - x_j}{x_i - x_j}. \qquad (8.69)
$$

For uniformly spaced samples

$$
x_i = x_0 + ih \qquad (8.70)
$$

and with the new variable α as given by

$$
x = x_0 + \alpha h, \qquad (8.71)
$$

we get

$$\frac{x - x_j}{x_i - x_j} = \frac{(x_0 + \alpha h) - (x_0 + jh)}{(x_0 + ih) - (x_0 + jh)} = \frac{\alpha - j}{i - j} \tag{8.72}$$

and hence

$$l_i(x(\alpha)) = \prod_{j=0, j \neq i}^{N} \frac{\alpha - j}{i - j}. \tag{8.73}$$

For even N we can write

$$l_i(x(\alpha)) = \prod_{j=-N/2, j \neq i}^{N/2} \frac{\alpha - j}{i - j}, \tag{8.74}$$

and for odd N,

$$l_i(x(\alpha)) = \prod_{j=-N-1/2, j \neq i}^{N+1/2} \frac{\alpha - j}{i - j}. \tag{8.75}$$

The interpolation of an output sample is given by

$$\boxed{y(n + \alpha) = \sum_{i=-N/2}^{N/2} l_i(\alpha) y(n + i).} \tag{8.76}$$

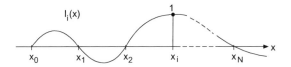

Figure 8.21 Lagrange polynomial.

Example: For $N = 2$, 3 samples,

$$l_{-1}(x(\alpha)) = \prod_{j=-1, j \neq -1}^{1} \frac{\alpha - j}{-1 - j} = \frac{1}{2}\alpha(\alpha - 1),$$

$$l_0(x(\alpha)) = \prod_{j=-1, j \neq 0}^{1} \frac{\alpha - j}{0 - j} = -(\alpha - 1)(\alpha + 1) = 1 - \alpha^2,$$

$$l_1(x(\alpha)) = \prod_{j=-1, j \neq 1}^{1} \frac{\alpha - j}{1 - j} = \frac{1}{2}\alpha(\alpha + 1).$$

8.4.3 Spline Interpolation

The interpolation using piecewise defined functions that only exist over finite intervals is called spline interpolation [Cuc91]. The goal is to compute the sample $y(n + \alpha) = \sum_{i=-N/2}^{N/2} b_i^N(\alpha) y(n + i)$ from weighted samples $y(n + i)$.

A B-spline $M_k^N(x)$ of Nth order using $m + 1$ samples is defined in the interval $[x_k, \ldots, x_{k+m}]$ by

$$M_k^N(x) = \sum_{i=k}^{k+m} a_i \phi_i(x) \tag{8.77}$$

with the truncated power functions

$$\phi_i(x) = (x - x_i)_+^N = \begin{cases} 0, & x < x_i, \\ (x - x_i)^N, & x \geq x_i. \end{cases} \tag{8.78}$$

In the following $M_0^N(x) = \sum_{i=0}^{m} a_i \phi_i(x)$ will be considered for $k = 0$ where $M_0^N(x) = 0$ for $x < x_0$ and $M_0^N(x) = 0$ for $x \geq x_m$. Figure 8.22 shows the truncated power functions and the B-spline of Nth order. With the definition of the truncated power functions we can write

$$\begin{aligned} M_0^N(x) &= a_0 \phi_0(x) + a_1 \phi_1(x) + \cdots + a_m \phi_m(x) \\ &= a_0(x - x_0)_+^N + a_1(x - x_1)_+^N + \cdots + a_m(x - x_m)_+^N, \end{aligned} \tag{8.79}$$

and after some calculations we get

$$\begin{aligned} M_0^N(x) &= a_0(x_0^N + c_1 x_0^{N-1} x + \cdots + c_{N-1} x_0 x^{N-1} + x^N) \\ &\quad + a_1(x_1^N + c_1 x_1^{N-1} x + \cdots + c_{N-1} x_1 x^{N-1} + x^N) \\ &\quad \vdots \\ &\quad + a_m(x_m^N + c_1 x_m^{N-1} x + \cdots + c_{N-1} x_m x^{N-1} + x^N). \end{aligned} \tag{8.80}$$

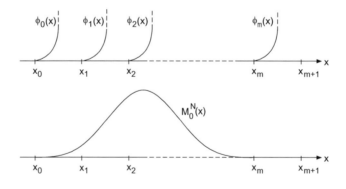

Figure 8.22 Truncated power functions and the B-spline of Nth order.

With the condition $M_0^N(x) = 0$ for $x \geq x_m$, the following set of linear equations can be written with (8.80) and the coefficients of the powers of x:

$$\begin{bmatrix} 1 & 1 & \cdots & 1 \\ x_0 & x_1 & \cdots & x_m \\ x_0^2 & x_1^2 & \cdots & x_m^2 \\ \vdots & \vdots & & \vdots \\ x_0^N & x_1^N & \cdots & x_m^N \end{bmatrix} \begin{bmatrix} a_0 \\ a_1 \\ a_2 \\ \vdots \\ a_m \end{bmatrix} = \begin{bmatrix} 0 \\ 0 \\ 0 \\ \vdots \\ 0 \end{bmatrix}. \tag{8.81}$$

The homogeneous set of linear equations has non-trivial solutions for $m > N$. The minimum requirement results in $m = N + 1$. For $m = N + 1$, the coefficients [Boe93] can be obtained as follows:

$$
a_i = \cfrac{\begin{vmatrix} 1 & 1 & 1 & \cdots & 0 & \cdots & 1 \\ x_0 & x_1 & x_2 & \cdots & 0 & \cdots & x_{N+1} \\ \vdots & \vdots & \vdots & & \vdots & & \vdots \\ x_0^N & x_1^N & x_2^N & \cdots & 0 & \cdots & x_{N+1}^N \end{vmatrix}}{\begin{vmatrix} 1 & 1 & 1 & \cdots & 1 \\ x_0 & x_1 & x_2 & \cdots & x_{N+1} \\ \vdots & \vdots & \vdots & & \vdots \\ x_0^{N+1} & x_1^{N+1} & x_2^{N+1} & \cdots & x_{N+1}^{N+1} \end{vmatrix}}, \quad i = 0, 1, \ldots, N + 1. \tag{8.82}
$$

ith column

Setting the ith column of the determinant in the numerator of (8.82) equal to zero corresponds to deleting the column. Computing both determinants of *Vandermonde* matrices [Bar90] and division leads to the coefficients

$$
a_i = \frac{1}{\prod_{j=0, i \neq j}^{N+1} (x_i - x_j)} \tag{8.83}
$$

and hence

$$
M_0^N(x) = \sum_{i=0}^{N+1} \frac{(x - x_i)_+^N}{\prod_{j=0, i \neq j}^{N+1} (x_i - x_j)}. \tag{8.84}
$$

For some k we obtain

$$
M_k^N(x) = \sum_{i=k}^{k+N+1} \frac{(x - x_i)_+^N}{\prod_{j=0, i \neq j}^{N+1} (x_i - x_j)}. \tag{8.85}
$$

Since the functions $M_k^N(x)$ decrease with increasing N, a normalization of the form $N_k^N(x) = (x_{k+N+1} - x_k) M_k^N$ is done, such that for equidistant samples we get

$$
N_k^N(x) = (N + 1) \cdot M_k^N(x). \tag{8.86}
$$

The next example illustrates the computation of B-splines.

Example: For $N = 3$, $m = 4$, and five samples the coefficients according to (8.83) are given by

$$a_0 = \frac{1}{(x_0 - x_4)(x_0 - x_3)(x_0 - x_2)(x_0 - x_1)},$$

$$a_1 = \frac{1}{(x_1 - x_4)(x_1 - x_3)(x_1 - x_2)(x_1 - x_0)},$$

$$a_2 = \frac{1}{(x_2 - x_4)(x_2 - x_3)(x_2 - x_1)(x_2 - x_0)},$$

$$a_3 = \frac{1}{(x_3 - x_4)(x_3 - x_2)(x_3 - x_1)(x_3 - x_0)},$$

$$a_4 = \frac{1}{(x_4 - x_3)(x_4 - x_2)(x_2 - x_1)(x_3 - x_0)}.$$

Figure 8.23a,b shows the truncated power functions and their summation for calculating $N_0^3(x)$. In Fig. 8.23c the horizontally shifted $N_i^3(x)$ are depicted.

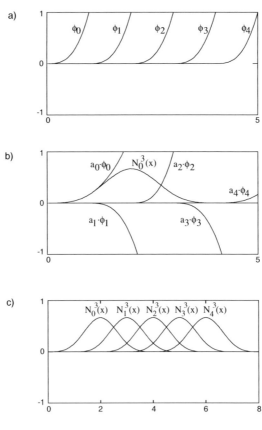

Figure 8.23 Third-order B-spline ($N = 3$, $m = 4$, 5 samples).

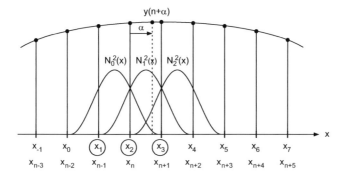

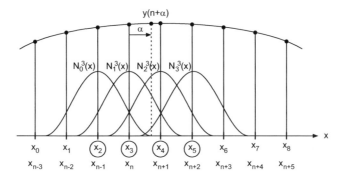

Figure 8.24 Interpolation with B-splines of second and third order.

A linear combination of B-splines is called a spline. Figure 8.24 shows the interpolation of sample $y(n + \alpha)$ for splines of second and third order. The shifted B-splines $N_i^N(x)$ are evaluated at the vertical line representing the distance α. With sample $y(n)$ and the normalized B-splines $N_i^N(x)$, the second- and third-order splines are respectively expressed as

$$y(n + \alpha) = \sum_{i=-1}^{1} y(n + i)N_{n-1+i}^2(\alpha) \qquad (8.87)$$

and

$$y(n + \alpha) = \sum_{i=-1}^{2} y(n + i)N_{n-2+i}^3(\alpha). \qquad (8.88)$$

The computation of a second-order B-spline at the sample index α is based on the symmetry properties of the B-spline, as depicted in Fig. 8.25. With (8.77), (8.86) and the symmetry properties shown in Fig. 8.25, the B-splines can be written in the form

$$N_2^2(\alpha) = N_0^2(\alpha) = 3 \sum_{i=0}^{3} a_i (\alpha - x_i)_+^2,$$

$$N_1^2(1 + \alpha) = N_0^2(1 + \alpha) = 3 \sum_{i=0}^{3} a_i (1 + \alpha - x_i)_+^2,$$

$$N_0^2(2 + \alpha) = N_0^2(1 - \alpha) = 3 \sum_{i=0}^{3} a_i (2 + \alpha - x_i)_+^2 = 3 \sum_{i=0}^{3} a_i (1 - \alpha - x_i)_+^2. \quad (8.89)$$

With (8.83) we get the coefficients

$$a_0 = \frac{1}{(0 - 1)(-2)(-3)} = -\frac{1}{6},$$

$$a_1 = \frac{1}{(1 - 0)(1 - 2)(1 - 3)} = \frac{1}{2},$$

$$a_2 = \frac{1}{(2 - 0)(2 - 1)(2 - 3)} = -\frac{1}{2}, \quad (8.90)$$

and thus

$$N_2^2(\alpha) = 3[a_0\alpha^2] = -\tfrac{1}{2}\alpha^2,$$

$$N_1^2(\alpha) = 3[a_0(1 + \alpha)^2 + a_1\alpha^2] = -\tfrac{1}{2}(1 + \alpha)^2 + \tfrac{3}{2}\alpha^2,$$

$$N_0^2(\alpha) = 3[a_0(1 - \alpha)^2] = -\tfrac{1}{2}(1 - \alpha)^2. \quad (8.91)$$

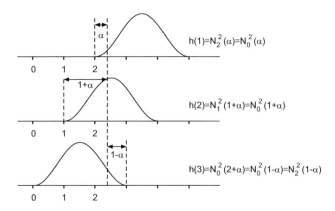

Figure 8.25 Exploiting the symmetry properties of a second-order B-spline.

Owing to the symmetrical properties of the B-splines, the time-variant coefficients of the second-order B-spline can be derived as

$$N_2^2(\alpha) = h(1) = -\tfrac{1}{2}\alpha^2, \quad (8.92)$$

$$N_1^2(\alpha) = h(2) = -\tfrac{1}{2}(1 + \alpha)^2 + \tfrac{3}{2}\alpha^2, \quad (8.93)$$

$$N_0^2(\alpha) = h(3) = -\tfrac{1}{2}(1 - \alpha)^2. \quad (8.94)$$

In the same way the time-variant coefficients of a third-order B-spline are given by

$$N_3^3(\alpha) = h(1) = \tfrac{1}{6}\alpha^3, \tag{8.95}$$

$$N_2^3(\alpha) = h(2) = \tfrac{1}{6}(1+\alpha)^3 - \tfrac{2}{3}\alpha^3, \tag{8.96}$$

$$N_1^3(\alpha) = h(3) = \tfrac{1}{6}(2-\alpha)^3 - \tfrac{2}{3}(1-\alpha)^3, \tag{8.97}$$

$$N_0^3(\alpha) = h(4) = \tfrac{1}{6}(1-\alpha)^3. \tag{8.98}$$

Higher-order B-splines are given by

$$y(n+\alpha) = \sum_{i=-2}^{2} y(n+i)N_{n-2+i}^4(\alpha), \tag{8.99}$$

$$y(n+\alpha) = \sum_{i=-2}^{3} y(n+i)N_{n-3+i}^5(\alpha), \tag{8.100}$$

$$y(n+\alpha) = \sum_{i=-3}^{3} y(n+i)N_{n-3+i}^6(\alpha). \tag{8.101}$$

Similar sets of coefficients can be derived here as well. Figure 8.26 illustrates this for fourth- and sixth-order B-splines.

Generally, for even orders we get

$$y(n+\alpha) = \sum_{i=-N/2}^{N/2} N_{N/2+i}^N(\alpha)y(n+i), \tag{8.102}$$

and for odd orders

$$y(n+\alpha) = \sum_{i=-(N-1)/2}^{(N+1)/2} N_{(N-1)/2+i}^N(\alpha)y(n+i). \tag{8.103}$$

For the application of interpolation the properties in the frequency domain are important. The zero-order B-spline is given by

$$N_0^0(x) = \sum_{i=0}^{1} a_i\phi_i(x) = \begin{cases} 0, & x < 0, \\ 1, & 0 \le x < 1, \\ 0, & x \ge 1, \end{cases} \tag{8.104}$$

and the Fourier transform gives the sinc function in the frequency domain. The first-order B-spline given by

$$N_0^1(x) = 2\sum_{i=0}^{2} a_i\phi_i(x) = \begin{cases} 0, & x < 0, \\ \dfrac{1}{2}x, & 0 \le x < 1, \\ 1 - \dfrac{1}{2}x, & 1 \le x < 2, \\ 0, & x \ge 2, \end{cases} \tag{8.105}$$

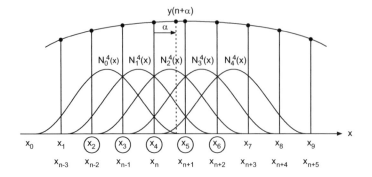

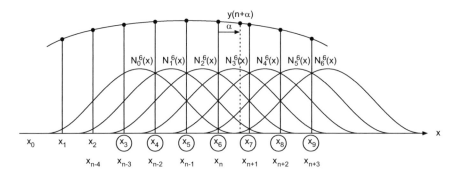

Figure 8.26 Interpolation with B-splines of fourth and sixth order.

leads to a sinc^2 function in the frequency domain. Higher-order B-splines can be derived by repeated convolution [Chu92] as given by

$$N^N(x) = N^0(x) * N^{N-1}(x). \tag{8.106}$$

Thus, the Fourier transform leads to

$$\mathrm{FT}[N^N(x)] = \mathrm{sinc}^{N+1}(f). \tag{8.107}$$

With the help of the properties in the frequency domain, the necessary order of the spline interpolation can be determined. Owing to the attenuation properties of the sinc$^{N+1}(f)$ function and the simple real-time calculation of the coefficients, spline interpolation is well suited to time-variant conversion in the last stage of a multistage sampling rate conversion system [Zöl94].

8.5 Exercises

1. Basics

Consider a simple sampling rate conversion system with a conversion rate of $\frac{4}{3}$. The system consists of two upsampling blocks, each by 2, and one downsampling block by 3.

1. What are anti-imaging and anti-aliasing filters and where do we need them in our system?

2. Sketch the block diagram.

3. Sketch the input, intermediate and output spectra in the frequency domain.

4. How is the amplitude affected by the up- and downsampling and where does it come from?

5. Sketch the frequency response of the anti-aliasing and anti-imaging filters needed for this upsampling system.

2. Synchronous Conversion

Our system will now be upsampled directly by a factor of 4, and again downsampled by a factor of 3, but with linear interpolation and decimation methods. The input signal is $x(n) = \sin(n\pi/6)$, $n = 0, \ldots, 48$.

1. What are the impulse responses of the two interpolation filters? Sketch their magnitude responses.

2. Plot the signals (input, intermediate and output signal) in the time domain using Matlab.

3. What is the delay resulting from the causal interpolation/decimation filters?

4. Show the error introduced by this interpolation/decimation method, in the frequency domain.

3. Polyphase Representation

Now we extend our system using a polyphase decomposition of the interpolation/decimation filters.

1. Sketch the idea of polyphase decomposition using a block diagram. What is the benefit of such decomposition?

2. Calculate the polyphase filters for up- and downsampling (using interpolation and decimation).

3. Use Matlab to plot all resulting signals in the time and frequency domain.

4. Asynchronous Conversion

1. What is the basic concept of asynchronous sampling rate conversion?

2. Sketch the block diagram and discuss the individual operations.

3. What is the necessary oversampling factor L for a 20-bit resolution?

4. How can we simplify the oversampling operations?

5. How can we make use of polyphase filtering?

6. Why are halfband filters an efficient choice for the upsampling operation?

7. Which parameters determine the interpolation algorithms in the last stage of the conversion?

References

[Ada92] R. Adams, T. Kwan: *VLSI Architectures for Asynchronous Sample-Rate Conversion*, Proc. 93rd AES Convention, Preprint No. 3355, San Francisco, October 1992.

[Ada93] R. Adams, T. Kwan: *Theory and VLSI Implementations for Asynchronous Sample-Rate Conversion*, Proc. 94th AES Convention, Preprint No. 3570, Berlin, March 1993.

[Bar90] S. Barnett: *Matrices – Methods and Applications*, Oxford University Press, Oxford, 1990.

[Boe93] W. Boehm, H. Prautzsch: *Numerical Methods*, Vieweg, Wiesbaden, 1993.

[Chu92] C. K. Chui, Ed.: *Wavelets: A Tutorial in Theory and Applications*, Volume 2, Academic Press, Boston, 1992.

[Cro83] R. E. Crochiere, L. R. Rabiner: *Multirate Digital Signal Processing*, Prentice Hall, Englewood Cliffs, NJ, 1983.

[Cuc91] S. Cucchi, F. Desinan, G. Parladori, G. Sicuranza: *DSP Implementation of Arbitrary Sampling Frequency Conversion for High Quality Sound Application*, Proc. IEEE ICASSP-91, pp. 3609–3612, Toronto, May 1991.

[Fli00] N. Fliege: *Multirate Digital Signal Processing*, John Wiley & Sons Ltd, Chichester, 2000.

[Hsi87] C.-C. Hsiao: *Polyphase Filter Matrix for Rational Sampling Rate Conversions*, Proc. IEEE ICASSP-87, pp. 2173–2176, Dallas, April 1987.

[Kat85] Y. Katsumata, O. Hamada: *A Digital Audio Sampling Frequency Converter Employing New Digital Signal Processors*, Proc. 79th AES Convention, Preprint No. 2272, New York, October 1985.

[Kat86]   Y. Katsumata, O. Hamada: *An Audio Sampling Frequency Conversion Using Digital Signal Processors*, Proc. IEEE ICASSP-86, pp. 33–36, Tokyo, 1986.

[Lag81]   R. Lagadec, H. O. Kunz: *A Universal, Digital Sampling Frequency Converter for Digital Audio*, Proc. IEEE ICASSP-81, pp. 595–598, Atlanta, April 1981.

[Lag82a]   R. Lagadec, D. Pelloni, D. Weiss: *A Two-Channel Professional Digital Audio Sampling Frequency Converter*, Proc. 71st AES Convention, Preprint No. 1882, Montreux, March 1982.

[Lag82b]   D. Lagadec, D. Pelloni, D. Weiss: *A 2-Channel, 16-Bit Digital Sampling Frequency Converter for Professional Digital Audio*, Proc. IEEE ICASSP-82, pp. 93–96, Paris, May 1982.

[Lag82c]   R. Lagadec: *Digital Sampling Frequency Conversion*, Digital Audio, Collected Papers from the AES Premier Conference, pp. 90–96, June 1982.

[Lag83]   R. Lagadec, D. Pelloni, A. Koch: *Single-Stage Sampling Frequency Conversion*, Proc. 74th AES Convention, Preprint No. 2039, New York, October 1983.

[Liu92]   G.-S. Liu, C.-H. Wei: *A New Variable Fractional Delay Filter with Nonlinear Interpolation*, IEEE Trans. Circuits and Systems-II: Analog and Digital Signal Processing, Vol. 39, No. 2, pp. 123–126, February 1992.

[Opp99]   A. V. Oppenheim, R. W. Schafer, J. R. Buck: *Discrete Time Signal Processing*, 2nd edn, Prentice Hall, Upper Saddle River, NJ, 1999.

[Par90]   S. Park, R. Robles: *A Real-Time Method for Sample-Rate Conversion from CD to DAT*, Proc. IEEE Int. Conf. Consumer Electronics, pp. 360–361, Chicago, June 1990.

[Par91a]   S. Park: *Low Cost Sample Rate Converters*, Proc. NAB Broadcast Engineering Conference, Las Vegas, April 1991.

[Par91b]   S. Park, R. Robles: *A Novel Structure for Real-Time Digital Sample-Rate Converters with Finite Precision Error Analysis*, Proc. IEEE ICASSP-91, pp. 3613–3616, Toronto, May 1991.

[Ram82]   T. A. Ramstad: *Sample-Rate Conversion by Arbitrary Ratios*, Proc. IEEE ICASSP-82, pp. 101–104, Paris, May 1982.

[Ram84]   T. A. Ramstad: *Digital Methods for Conversion between Arbitrary Sampling Frequencies*, IEEE Trans. on Acoustics, Speech and Signal Processing, Vol. ASSP-32, No. 3, pp. 577–591, June 1984.

[Smi84]   J. O. Smith, P. Gossett: *A Flexible Sampling-Rate Conversion Method*, Proc. IEEE ICASSP-84, pp. 19.4.1–19.4.4, 1984.

[Sti91]   E. F. Stikvoort: *Digital Sampling Rate Converter with Interpolation in Continuous Time*, Proc. 90th AES Convention, Preprint No. 3018, Paris, February 1991.

[Vai93]   P. P. Vaidyanathan: *Multirate Systems and Filter Banks*, Prentice Hall, Englewood Cliffs, NJ, 1993.

[Zöl94]   U. Zölzer, T. Boltze: *Interpolation Algorithms: Theory and Application*, Proc. 97th AES Convention, Preprint No. 3898, San Francisco, November 1994.

Chapter 9

Audio Coding

For transmission and storage of audio signals, different methods for compressing data have been investigated besides the pulse code modulation representation. The requirements of different applications have led to a variety of audio coding methods which have become international standards. In this chapter basic principles of audio coding are introduced and the most important audio coding standards discussed. Audio coding can be divided into two types: *lossless* and *lossy*. Lossless audio coding is based on a statistical model of the signal amplitudes and coding of the audio signal (audio coder). The reconstruction of the audio signal at the receiver allows a lossless resynthesis of the signal amplitudes of the original audio signal (audio decoder). On the other hand, lossy audio coding makes use of a psychoacoustic model of human acoustic perception to quantize and code the audio signal. In this case only the acoustically relevant parts of the signal are coded and reconstructed at the receiver. The samples of the original audio signal are not exactly reconstructed. The objective of both audio coding methods is a data rate reduction or data compression for transmission or storage compared to the original PCM signal.

9.1 Lossless Audio Coding

Lossless audio coding is based on linear prediction followed by entropy coding [Jay84] as shown in Fig. 9.1:

- *Linear Prediction.* A quantized set of coefficients P for a block of M samples is determined which leads to an estimate $\hat{x}(n)$ of the input sequence $x(n)$. The aim is to minimize the power of the difference signal $d(n)$ without any additional quantization errors, i.e. the word-length of the signal $\hat{x}(n)$ must be equal to the word-length of the input. An alternative approach [Han98, Han01] quantizes the prediction signal $\hat{x}(n)$ such that the word-length of the difference signals $d(n)$ remains the same as the input signal word-length. Figure 9.2 shows a signal block $x(n)$ and the corresponding spectrum $|X(f)|$. Filtering the input signal with the predictor filter transfer function $P(z)$ delivers the estimate $\hat{x}(n)$. Subtracting input and prediction signal yields the prediction error $d(n)$, which is also shown in Fig. 9.2 and which has a considerably lower power compared to the input power. The spectrum of this prediction error is

Digital Audio Signal Processing   Second Edition   Udo Zölzer
© 2008 John Wiley & Sons, Ltd

nearly white (see Fig. 9.2, lower right). The prediction can be represented as a filter operation with an analysis transfer function $H_A(z) = 1 - P(z)$ on the coder side.

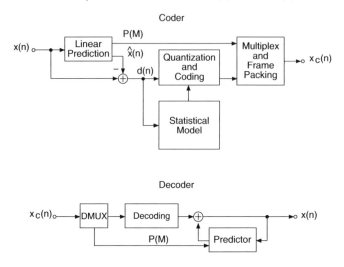

Figure 9.1 Lossless audio coding based on linear prediction and entropy coding.

- *Entropy Coding.* Quantization of signal $d(n)$ due to the probability density function of the block. Samples $d(n)$ of greater probability are coded with shorter data words, whereas samples $d(n)$ of lesser probability are coded with longer data words [Huf52].

- *Frame Packing.* The frame packing uses the quantized and coded difference signal and the coding of the M coefficients of the predictor filter $P(z)$ of order M.

- *Decoder.* On the decoder side the inverse synthesis transfer function $H_S(z) = H_A^{-1}(z) = [1 - P(z)]^{-1}$ reconstructs the input signal with the coded difference samples and the M filter coefficients. The frequency response of this synthesis filter represents the spectral envelope shown in the upper right part of Fig. 9.2. The synthesis filter shapes the white spectrum of the difference (prediction error) signal with the spectral envelope of the input spectrum.

The attainable compression rates depend on the statistics of the audio signal and allow a compression rate of up to 2 [Bra92, Cel93, Rob94, Cra96, Cra97, Pur97, Han98, Han01, Lie02, Raa02, Sch02]. Figure 9.3 illustrates examples of the necessary word-length for lossless audio coding [Blo95, Sqa88]. Besides the local entropy of the signal (entropy computed over a block length of 256), results for linear prediction followed by Huffman coding [Huf52] are presented. Huffman coding is carried out with a fixed code table [Pen93] and a power-controlled choice of adapted code tables. It is observed from Fig. 9.3 that for high signal powers, a reduction in word-length is possible if the choice is made from several adapted code tables. Lossless compression methods are used for storage media with limited word-length (16 bits in CD and DAT) which are used for recording audio signals of higher word-lengths (more than 16 bits). Further applications are in the transmission and archiving of audio signals.

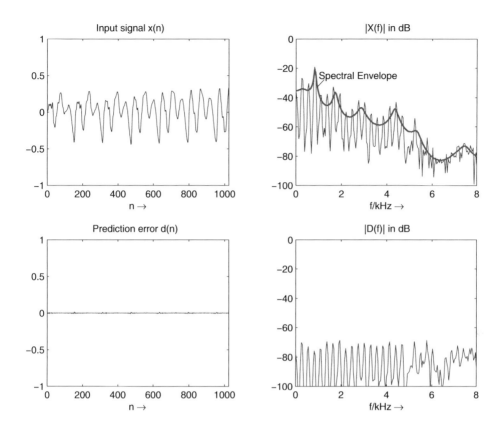

Figure 9.2 Signals and spectra for linear prediction.

9.2 Lossy Audio Coding

Significantly higher compression rates (of factor 4 to 8) can be obtained with *lossy* coding methods. Psychoacoustic phenomena of human hearing are used for signal compression. The fields of application have a wide range, from professional audio like source coding for DAB to audio transmission via ISDN and home entertainment like DCC and MiniDisc.

An outline of the coding methods [Bra94] is standardized in an international specification ISO/IEC 11172-3 [ISO92], which is based on the following processing (see Fig. 9.4):

- subband decomposition with filter banks of short latency time;

- calculation of psychoacoustic model parameters based on short-time FFT;

- dynamic bit allocation due to psychoacoustic model parameters (signal-to-mask ratio SMR);

- quantization and coding of subband signals;

- multiplex and frame packing.

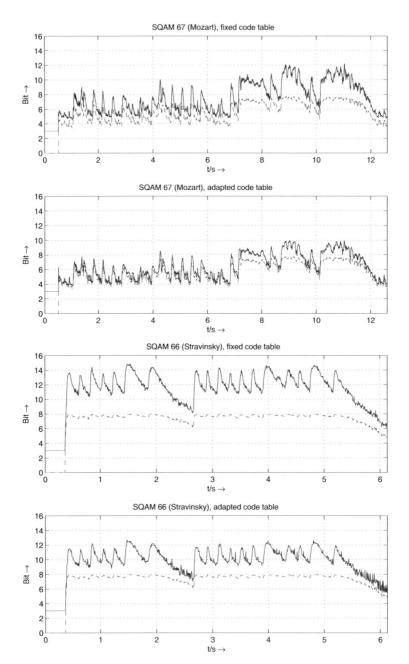

Figure 9.3 Lossless audio coding (Mozart, Stravinsky): word-length in bits versus time (entropy - -, linear prediction with Huffman coding —).

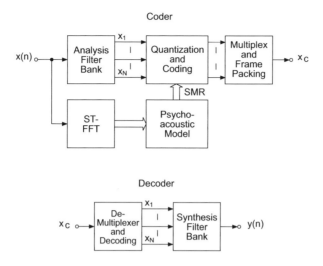

Figure 9.4 Lossy audio coding based on subband coding and psychoacoustic models.

Owing to lossy audio coding, post-processing of such signals or several coding and decoding steps is associated with some additional problems. The high compression rates justify the use of lossy audio coding techniques in applications like transmission.

9.3 Psychoacoustics

In this section, basic principles of psychoacoustics are presented. The results of psychoacoustic investigations by Zwicker [Zwi82, Zwi90] form the basis for audio coding based on models of human perception. These coded audio signals have a significantly reduced data rate compared to the linearly quantized PCM representation. The human auditory system analyzes broad-band signals in so-called critical bands. The aim of psychoacoustic coding of audio signals is to decompose the broad-band audio signal into subbands which are matched to the critical bands and then perform quantization and coding of these subband signals [Joh88a, Joh88b, Thei88]. Since the perception of sound below the absolute threshold of hearing is not possible, subband signals below this threshold need neither be coded nor transmitted. In addition to the perception in critical bands and the absolute threshold, the effects of signal masking in human perception play an important role in signal coding. These are explained in the following and their application to psychoacoustic coding is discussed.

9.3.1 Critical Bands and Absolute Threshold

Critical Bands. Critical bands as investigated by Zwicker are listed in Table 9.1.

Table 9.1 Critical bands as given by Zwicker [Zwi82].

z/Bark	f_l/Hz	f_u/Hz	f_B/Hz	f_c/Hz
0	0	100	100	50
1	100	200	100	150
2	200	300	100	250
3	300	400	100	350
4	400	510	110	450
5	510	630	120	570
6	630	770	140	700
7	770	920	150	840
8	920	1080	160	1000
9	1080	1270	190	1170
10	1270	1480	210	1370
11	1480	1720	240	1600
12	1720	2000	280	1850
13	2000	2320	320	2150
14	2320	2700	380	2500
15	2700	3150	450	2900
16	3150	3700	550	3400
17	3700	4400	700	4000
18	4400	5300	900	4800
19	5300	6400	1100	5800
20	6400	7700	1300	7000
21	7700	9500	1800	8500
22	9500	1200	2500	10500
23	12000	15500	3500	13500
24	15500			

A transformation of the linear frequency scale into a hearing-adapted scale is given by Zwicker [Zwi90] (units of z in Bark):

$$\frac{z}{\text{Bark}} = 13 \arctan\left(0.76\frac{f}{\text{kHz}}\right) + 3.5 \arctan\left(\frac{f}{7.5\ \text{kHz}}\right)^2. \tag{9.1}$$

The individual critical bands have bandwidths

$$\Delta f_B = 25 + 75\left(1 + 1.4\left(\frac{f}{\text{kHz}}\right)^2\right)^{0.69}. \tag{9.2}$$

Absolute Threshold. The absolute threshold L_{T_q} (threshold in quiet) denotes the curve of sound pressure level L [Zwi82] versus frequency, which leads to the perception of a sinusoidal tone. The absolute threshold is given by [Ter79]:

$$\frac{L_{T_q}}{\text{dB}} = 3.64\left(\frac{f}{\text{kHz}}\right)^{-0.8} - 6.5 \exp\left(-0.6\left(\frac{f}{\text{kHz}} - 3.3\right)^2\right) + 10^{-3}\left(\frac{f}{\text{kHz}}\right)^4. \tag{9.3}$$

Below the absolute threshold, no perception of signals is possible. Figure 9.5 shows the absolute threshold versus frequency. Band-splitting in critical bands and the absolute threshold allow the calculation of an offset between the signal level and the absolute threshold for every critical band. This offset is responsible for choosing appropriate quantization steps for each critical band.

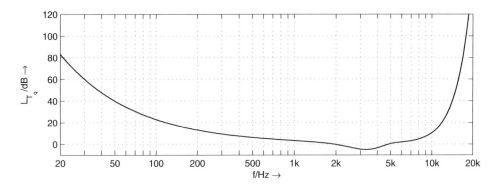

Figure 9.5 Absolute threshold (threshold in quiet).

9.3.2 Masking

For audio coding the use of sound perception in critical bands and absolute threshold only is not sufficient for high compression rates. The basis for further data reduction are the masking effects investigated by Zwicker [Zwi82, Zwi90]. For band-limited noise or a sinusoidal signal, frequency-dependent masking thresholds can be given. These thresholds perform masking of frequency components if these components are below a masking threshold (see Fig. 9.6). The application of masking for perceptual coding is described in the following.

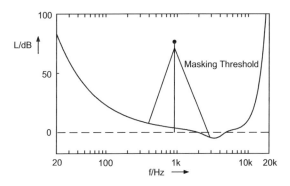

Figure 9.6 Masking threshold of band-limited noise.

Calculation of Signal Power in Band i. First, the sound pressure level within a critical band is calculated. The short-time spectrum $X(k) = \text{DFT}[x(n)]$ is used to calculate the power density spectrum

$$S_p(e^{j\Omega}) = S_p(e^{j(2\pi k/N)}) = X_R^2(e^{j(2\pi k/N)}) + X_I^2(e^{j(2\pi k/N)}), \tag{9.4}$$

$$S_p(k) = X_R^2(k) + X_I^2(k), \quad 0 \le k \le N-1, \tag{9.5}$$

with the help of an N-point FFT. The signal power in band i is calculated by the sum

$$S_p(i) = \sum_{\Omega=\Omega_{li}}^{\Omega_{ui}} S_p(k) \qquad (9.6)$$

from the lower frequency up to the upper frequency of critical band i. The sound pressure level in band i is given by $L_S(i) = 10 \log_{10} S_p(i)$.

Absolute Threshold. The absolute threshold is set such that a 4 kHz signal with peak amplitude ± 1 LSB for a 16-bit representation lies at the lower limit of the absolute threshold curve. Every masking threshold calculated in individual critical bands, which lies below the absolute threshold, is set to a value equal to the absolute threshold in the corresponding band. Since the absolute threshold within a critical band varies for low and high frequencies, it is necessary to make use of the mean absolute threshold within a band.

Masking Threshold. The offset between signal level and the masking threshold in critical band i (see Fig. 9.7) is given by [Hel72]

$$\frac{O(i)}{\text{dB}} = \alpha(14.5 + i) + (1 - \alpha)a_v, \qquad (9.7)$$

where α denotes the tonality index and a_v is the masking index. The masking index [Kap92] is given by

$$a_v = -2 - 2.05 \arctan\left(\frac{f}{4 \text{ kHz}}\right) - 0.75 \arctan\left(\frac{f^2}{2.56 \text{ kHz}^2}\right). \qquad (9.8)$$

As an approximation,

$$\frac{O(i)}{\text{dB}} = \alpha(14.5 + i) + (1 - \alpha)5.5 \qquad (9.9)$$

can be used [Joh88a, Joh88b]. If a tone is masking a noise-like signal ($\alpha = 1$), the threshold is set $14.5 + i$ dB below the value of $L_S(i)$. If a noise-like signal is masking a tone ($\alpha = 0$), the threshold is set $5.5 + i$ dB below $L_S(i)$. In order to recognize a tonal or noise-like signal within a certain number of samples, the *spectral flatness measure* SFM is estimated. The SFM is defined by the ratio of the geometric to arithmetic mean value of $S_p(i)$ as

$$\text{SFM} = 10 \log_{10} \left(\frac{\left[\prod_{k=1}^{N/2} S_p(e^{j(2\pi k/N)}) \right]^{1/(N/2)}}{\frac{1}{N/2} \sum_{k=1}^{N/2} S_p(e^{j(2\pi k/N)})} i \right). \qquad (9.10)$$

The SFM is compared with the SFM of a sinusoidal signal (definition $\text{SFM}_{\text{max}} = -60$ dB) and the tonality index is calculated [Joh88a, Joh88b] by

$$\alpha = \min\left(\frac{\text{SFM}}{\text{SFM}_{\text{max}}}, 1 \right). \qquad (9.11)$$

SFM $= 0$ dB corresponds to a noise-like signal and leads to $\alpha = 0$, whereas an SFM$=75$ dB gives a tone-like signal ($\alpha = 1$). With the sound pressure level $L_S(i)$ and the offset $O(i)$

the masking threshold is given by

$$T(i) = 10^{[L_S(i)-O(i)]/10}. \tag{9.12}$$

Masking across Critical Bands. Masking across critical bands can be carried out with the help of the Bark scale. The masking threshold is of a triangular form which decreases at S_1 dB per Bark for the lower slope and at S_2 dB per Bark for the upper slope, depending on the sound pressure level L_i and the center frequency f_{c_i} in band i (see [Ter79]) according to

$$S_1 = 27 \text{ dB/Bark}, \tag{9.13}$$

$$S_2 = 24 + 0.23\left(\frac{f_{c_i}}{\text{kHz}}\right)^{-1} - 0.2\frac{L_S(i)}{\text{dB}} \text{ dB/Bark}. \tag{9.14}$$

An approximation of the minimum masking within a critical band can be made using Fig. 9.8 [Thei88, Sauv90]. Masking at the upper frequency f_{u_i} in the critical band i is responsible for masking the quantization noise with approximately 32 dB using the lower masking threshold that decreases by 27 dB/Bark. The upper slope has a steepness which depends on the sound pressure level. This steepness is lower than the steepness of the lower slope. Masking across critical bands is presented in Fig. 9.9. The masking signal in critical band $i - 1$ is responsible for masking the quantization noise in critical band i as well as the masking signal in critical band i. This kind of masking across critical bands further reduces the number of quantization steps within critical bands.

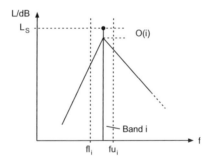

Figure 9.7 Offset between signal level and masking threshold.

An analytical expression for masking across critical bands [Schr79] is given by

$$10 \log_{10}[B(\Delta i)] = 15.81 + 7.5(\Delta i + 0.474) - 17.5[1 + (\Delta i + 0.474)^2]^{\frac{1}{2}}. \tag{9.15}$$

Δi denotes the distance between two critical bands in Bark. Expression (9.15) is called the *spreading* function. With the help of this *spreading* function, masking of critical band i by critical band j can be calculated [Joh88a, Joh88b] with $\text{abs}(i - j) \leq 25$ such that

$$S_m(i) = \sum_{j=0}^{24} B(i - j) \cdot S_p(i). \tag{9.16}$$

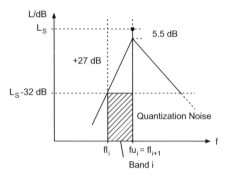

Figure 9.8 Masking within a critical band.

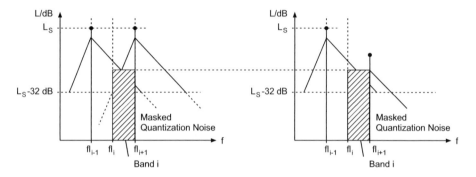

Figure 9.9 Masking across critical bands.

The masking across critical bands can therefore be expressed as a matrix operation given by

$$
\begin{bmatrix} S_m(0) \\ S_m(1) \\ \vdots \\ S_m(24) \end{bmatrix} = \begin{bmatrix} B(0) & B(-1) & B(-2) & \cdots & B(-24) \\ B(1) & B(0) & B(-1) & \cdots & B(-23) \\ \vdots & \vdots & \vdots & & \vdots \\ B(24) & B(23) & B(22) & \cdots & B(0) \end{bmatrix} \begin{bmatrix} S_p(0) \\ S_p(1) \\ \vdots \\ S_p(24) \end{bmatrix}. \tag{9.17}
$$

A renewed calculation of the masking threshold with (9.16) leads to the global masking threshold

$$
T_m(i) = 10^{\log_{10} S_m(i) - O(i)/10}. \tag{9.18}
$$

For a clarification of the single steps for a psychoacoustic based audio coding we summarize the operations with exemplified analysis results:

- calculation of the signal power $S_p(i)$ in critical bands $\rightarrow L_S(i)$ in dB (Fig. 9.10a);

- calculation of masking across critical bands $T_m(i) \rightarrow L_{T_m}(i)$ in dB (Fig. 9.10b);

- masking with tonality index $\rightarrow L_{T_m}(i)$ in dB (Fig. 9.10c);

- calculation of global masking threshold with respect to threshold in quiet $L_{T_q} \rightarrow L_{T_{m,\mathrm{abs}}}(i)$ in dB (Fig. 9.10d).

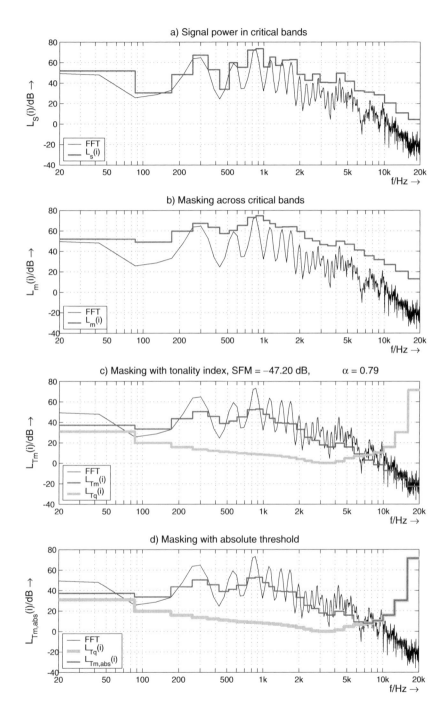

Figure 9.10 Stepwise calculation of psychoacoustic model.

With the help of the global masking threshold $L_{T_m,\text{abs}}(i)$ we calculate the signal-to-mask ratio

$$\text{SMR}(i) = L_S(i) - L_{T_m,\text{abs}}(i) \quad \text{in dB} \tag{9.19}$$

per Bark band. This signal-to-mask ratio defines the necessary number of bits per critical band, such that masking of quantization noise is achieved. For the given example the signal power and the global masking threshold are shown in Fig. 9.11a. The resulting signal-to-mask ratio $\text{SMR}(i)$ is shown in Fig. 9.11b. As soon as $\text{SMR}(i) > 0$, one has to allocate bits to the critical band i. For $\text{SMR}(i) < 0$ the corresponding critical band will not be transmitted. Figure 9.12 shows the masking thresholds in critical bands for a sinusoid of 440 Hz. Compared to the first example, the influence of masking thresholds across critical bands is easier to observe and interpret.

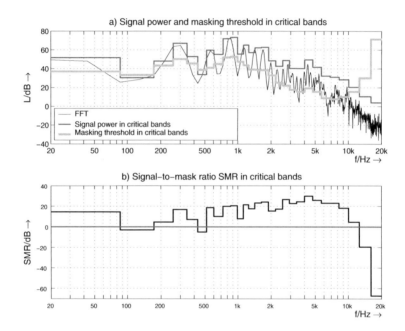

Figure 9.11 Calculation of the signal-to-mask ratio SMR.

9.4   ISO-MPEG-1 Audio Coding

In this section, the coding method for digital audio signals is described which is specified in the standard ISO/IEC 11172-3 [ISO92]. The filter banks used for subband decomposition, the psychoacoustic models, dynamic bit allocation and coding are discussed. A simplified block diagram of the coder for implementing layers I and II of the standard is shown in Fig. 9.13. The corresponding decoder is shown in Fig. 9.14. It uses the information from the ISO-MPEG1 frame and feeds the decoded subband signals to a synthesis filter bank for reconstructing the broad-band PCM signal. The complexity of the decoder is, in contrast

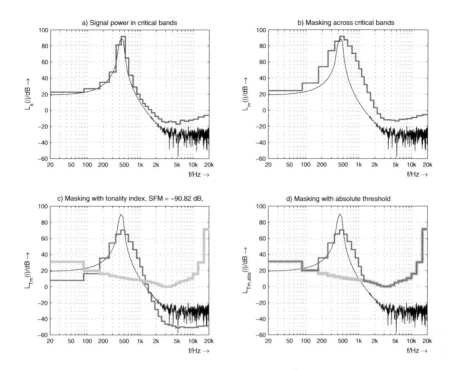

Figure 9.12 Calculation of psychoacoustic model for a pure sinusoid with 440 Hz.

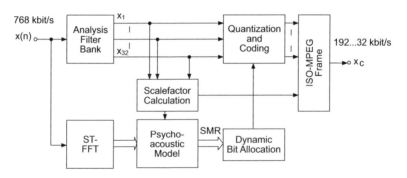

Figure 9.13 Simplified block diagram of an ISO-MPEG1 coder.

to the coder, significantly lower. Prospective improvements of the coding method are being made entirely for the coder.

9.4.1 Filter Banks

The subband decomposition is done with a pseudo-QMF filter bank (see Fig. 9.15). The theoretical background is found in the related literature [Rot83, Mas85, Vai93]. The broad-

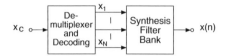

Figure 9.14 Simplified block diagram of an ISO-MPEG1 decoder.

band signal is decomposed into M uniformly spaced subbands. The subbands are processed further after a sampling rate reduction by a factor of M. The implementation of an ISO-MPEG1 coder is based on $M = 32$ frequency bands. The individual band-pass filters $H_0(z) \cdots H_{M-1}(z)$ are designed using a prototype low-pass filter $H(z)$ and frequency-shifted versions. The frequency shifting of the prototype with cutoff frequency $\pi/2M$ is done by modulating the impulse response $h(n)$ with a cosine term [Bos02] according to

$$h_k(n) = h(n) \cdot \cos\left(\frac{\pi}{32}(k + 0{,}5)(n - 16)\right), \tag{9.20}$$

$$f_k(n) = 32 \cdot h(n) \cdot \cos\left(\frac{\pi}{32}(k + 0{,}5)(n + 16)\right), \tag{9.21}$$

with $k = 0, \ldots, 31$ and $n = 0, \ldots, 511$. The band-pass filters have bandwidth π/M. For the synthesis filter bank, corresponding filters $F_0(z) \cdots F_{M-1}(z)$ give outputs which are added together, resulting in a broad-band PCM signal. The prototype impulse response with 512 taps, the modulated band-pass impulse responses, and the corresponding magnitude responses are shown in Fig. 9.16. The magnitude responses of all 32 band-pass filters are also shown. The overlap of neighboring band-pass filters is limited to the lower and upper filter band. This overlap reaches up to the center frequency of the neighboring bands. The resulting aliasing after downsampling in each subband will be canceled in the synthesis filter bank. The pseudo-QMF filter bank can be implemented by the combination of a polyphase filter structure followed by a discrete cosine transform [Rot83, Vai93, Kon94].

To increase the frequency resolution, layer III of the standard decomposes each of the 32 subbands further into a maximum of 18 uniformly spaced subbands (see Fig. 9.17). The decomposition is carried out with the help of an overlapped transform of windowed subband samples. The method is based on a modified discrete cosine transform, also known as the TDAC filter bank (*Time Domain Aliasing Cancellation*) and MLT (*Modulated Lapped Transform*). An exact description is given in [Pri87, Mal92]. This extended filter bank is referred to as the polyphase/modified discrete cosine transform (MDCT) hybrid filter bank [Bra94]. The higher frequency resolution enables a higher coding gain but has the disadvantage of a worse time resolution. This is observed for impulse-like signals. In order to minimize these artifacts, the number of subbands per subband can be altered from 18 down to 6. Subband decompositions that are matched to the signal can be obtained by specially designed window functions with overlapping transforms [Edl89, Edl95]. The equivalence of overlapped transforms and filter banks is found in [Mal92, Glu93, Vai93, Edl95, Vet95].

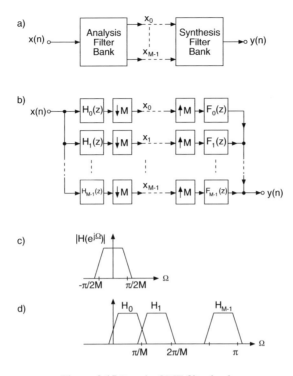

Figure 9.15 Pseudo-QMF filter bank.

9.4.2 Psychoacoustic Models

Two psychoacoustic models have been developed for layers I to III of the ISO-MPEG1 standard. Both models can be used independently of each other for all three layers. Psychoacoustic model 1 is used for layers I and II, whereas model 2 is used for layer III. Owing to the numerous applications of layers I and II, we discuss psychoacoustic model 1 in this subsection.

Bit allocation in each of the 32 subbands is carried out using the signal-to-mask ratio SMR(i). This is based on the minimum masking threshold and the maximum signal level within a subband. In order to calculate this ratio, the power density spectrum is estimated with the help of a short-time FFT in parallel with the analysis filter bank. As a consequence, a higher frequency resolution is obtained for estimating the power density spectrum in contrast to the frequency resolution of the 32-band analysis filter bank. The signal-to-mask ratio for every subband is determined as follows:

1. Calculate the power density spectrum of a block of N samples using FFT. After windowing a block of $N = 512$ ($N = 1024$ for layer II) input samples, the power density spectrum

$$X(k) = 10 \log_{10} \left| \frac{1}{N} \sum_{n=0}^{N-1} h(n)x(n)e^{-jnk2\pi/N} \right|^2 \quad \text{in dB} \qquad (9.22)$$

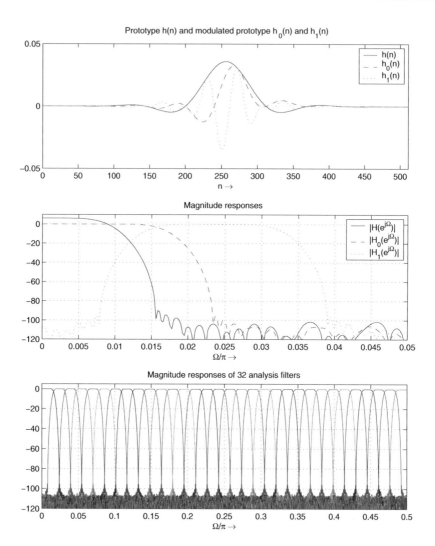

Figure 9.16 Impulse responses and magnitude responses of pseudo-QMF filter bank.

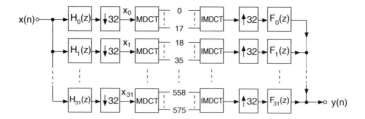

Figure 9.17 Polyphase/MDCT hybrid filter bank.

is calculated. Then the window $h(n)$ is displaced by 384 (12 · 32) samples and the next block is processed.

2. Determine the sound pressure level in every subband. The sound pressure level is derived from the calculated power density spectrum and by calculating a scaling factor in the corresponding subband as given by

$$L_S(i) = \max[X(k),\ 20\log_{10}[\text{SCF}_{\max}(i) \cdot 32768] - 10] \quad \text{in dB.} \qquad (9.23)$$

For $X(k)$, the maximum of the spectral lines in a subband is used. The scaling factor $\text{SCF}(i)$ for subband i is calculated from the absolute value of the maximum of 12 consecutive subband samples. A nonlinear quantization to 64 levels is carried out (layer I). For layer II, the sound pressure level is determined by choosing the largest of the three scaling factors from 3 · 12 subband samples.

3. Consider the absolute threshold. The absolute threshold $LT_q(m)$ is specified for different sampling rates in [ISO92]. The frequency index m is based on a reduction of $N/2$ relevant frequencies with the FFT of index k (see Fig. 9.18). The subband index is still i.

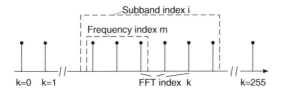

Figure 9.18 Nomenclature of frequency indices.

4. Calculate tonal $X_{tm}(k)$ or non-tonal $X_{nm}(k)$ masking components and determining relevant masking components (for details see [ISO92]). These masking components are denoted by $X_{tm}[z(j)]$ and $X_{nm}[z(j)]$. With the index j, tonal and non-tonal masking components are labeled. The variable $z(m)$ is listed for reduced frequency indices m in [ISO92]. It allows a finer resolution of the 24 critical bands with the frequency group index z.

5. Calculate the individual masking thresholds. For masking thresholds of tonal and non-tonal masking components $X_{tm}[z(j)]$ and $X_{nm}[z(j)]$, the following calculation is performed:

$$LT_{tm}[z(j), z(m)] = X_{tm}[z(j)] + a_{v_{tm}}[z(j)] + v_f[z(j), z(m)] \text{ dB}, \qquad (9.24)$$
$$LT_{nm}[z(j), z(m)] = X_{nm}[z(j)] + a_{v_{nm}}[z(j)] + v_f[z(j), z(m)] \text{ dB}. \qquad (9.25)$$

The masking index for tonal masking components is given by

$$a_{v_{tm}} = -1.525 - 0.275 \cdot z(j) - 4.5 \quad \text{in dB}, \qquad (9.26)$$

and the masking index for non-tonal masking components is

$$a_{v_{nm}} = -1.525 - 0.175 \cdot z(j) - 0.5 \quad \text{in dB}. \qquad (9.27)$$

The masking function $v_f[z(j), z(m)]$ with distance $\Delta z = z(m) - z(j)$ is given by

$$v_f = \begin{cases} 17 \cdot (\Delta z + 1) - (0.4 \cdot X[z(j)] + 6) & -3 \le \Delta z < -1 \\ (0.4 \cdot X[z(j)] + 6) \cdot \Delta z & -1 \le \Delta z < 0 \\ -17 \cdot \Delta z & 0 \le \Delta z < 1 \\ -(\Delta z - 1) \cdot (17 - 0.15 \cdot X[z(j)]) - 17 & 1 \le \Delta z < 8 \\ \text{in dB} & \text{in Bark.} \end{cases}$$

This masking function $v_f[z(j), z(m)]$ describes the masking of the frequency index $z(m)$ by the masking component $z(j)$.

6. Calculate the global masking threshold. For frequency index m, the global masking threshold is calculated as the sum of all contributing masking components according to

$$LT_g(m) = 10 \log_{10} \left[10^{LT_q(m)/10} + \sum_{j=1}^{T_m} 10^{LT_{tm}[z(j),z(m)]/10} \right.$$
$$\left. + \sum_{j=1}^{R_m} 10^{LT_{nm}[z(j),z(m)]/10} \right] \text{dB}. \tag{9.28}$$

The total number of tonal and non-tonal masking components are denoted as T_m and R_m respectively. For a given subband i, only masking components that lie in the range -8 to $+3$ Bark will be considered. Masking components outside this range are neglected.

7. Determine the minimum masking threshold in every subband:

$$LT_{\min}(i) = \min[LT_g(m)] \text{ dB}. \tag{9.29}$$

Several masking thresholds $LT_g(m)$ can occur in a subband as long as m lies within the subband i.

8. Calculate the signal-to-mask ratio $SMR(i)$ in every subband:

$$SMR(i) = L_S(i) - LT_{\min}(i) \text{ dB}. \tag{9.30}$$

The signal-to-mask ratio determines the dynamic range that has to be quantized in the particular subband so that the level of quantization noise lies below the masking threshold. The signal-to-mask ratio is the basis for the bit allocation procedure for quantizing the subband signals.

9.4.3 Dynamic Bit Allocation and Coding

Dynamic Bit Allocation. Dynamic bit allocation is used to determine the number of bits that are necessary for the individual subbands so that a transparent perception is possible. The minimum number of bits in subband i can be determined from the difference between scaling factor $SCF(i)$ and the absolute threshold $LT_q(i)$ as $b(i) = SCF(i) - LT_q(i)$. With this, quantization noise remains under the masking threshold. Masking across critical bands is used for the implementation of the ISO-MPEG1 coding method.

For a given transmission rate, the maximum possible number of bits B_m for coding subband signals and scaling factors is calculated as

$$B_m = \sum_{i=1}^{32} b(i) + \text{SCF}(i) + \text{additional information.} \qquad (9.31)$$

The bit allocation is performed within an allocation frame consisting of 12 subband samples ($384 = 12 \cdot 32$ PCM samples) for layer I and 36 subband samples ($1152 = 36 \cdot 32$ PCM samples) for layer II.

The dynamic bit allocation for the subband signals is carried out as an iterative procedure. At the beginning, the number of bits per subband is set to zero. First, the mask-to-noise ratio

$$\text{MNR}(i) = \text{SNR}(i) - \text{SMR}(i) \qquad (9.32)$$

is determined for every subband. The signal-to-mask ratio $\text{SMR}(i)$ is the result of the psychoacoustic model. The signal-to-noise ratio $\text{SNR}(i)$ is defined by a table in [ISO92], in which for every number of bits a corresponding signal-to-noise ratio is specified. The number of bits must be increased as long as the mask-to-noise ratio MNR is less than zero.

The iterative bit allocation is performed by the following steps.

1. Determination of the minimum $\text{MNR}(i)$ of all subbands.

2. Increasing the number of bits of these subbands on to the next stage of the MPEG1 standard. Allocation of 6 bits for the scaling factor of the MPEG1 standard when the number of bits is increased for the first time.

3. New calculation of $\text{MNR}(i)$ in this subband.

4. Calculation of the number of bits for all subbands and scaling factors and comparison with the maximum number B_m. If the number of bits is smaller than the maximum number, the iteration starts again with step 1.

Quantization and Coding of Subband Signals. The quantization of the subband signals is done with the allocated bits for the corresponding subband. The 12 (36) subband samples are divided by the corresponding scaling factor and then linearly quantized and coded (for details see [ISO92]). This is followed by a frame packing. In the decoder, the procedure is reversed. The decoded subband signals with different word-lengths are reconstructed into a broad-band PCM signal with a synthesis filter bank (see Fig. 9.14). MPEG-1 audio coding has a one- or a two-channel stereo mode with sampling frequencies of 32, 44.1, and 48 kHz and a bit rate of 128 kbit/s per channel.

9.5 MPEG-2 Audio Coding

The aim of the introduction of MPEG-2 audio coding was the extension of MPEG-1 to lower sampling frequencies and multichannel coding [Bos97]. Backward compatibility to existing MPEG-1 systems is achieved through the version MPEG-2 BC (Backward Compatible) and the introduction toward lower sampling frequencies of 32, 22.05, 24 kHz with version MPEG-2 LSF (Lower Sampling Frequencies). The bit rate for a five-channel MPEG-2 BC coding with full bandwidth of all channels is 640–896 kBit/s.

9.6 MPEG-2 Advanced Audio Coding

To improve the coding of mono, stereo, and multichannel audio signals the MPEG-2 AAC (Advanced Audio Coding) standard was specified. This coding standard is not backward compatible with the MPEG-1 standard and forms the kernel for new extended coding standards such as MPEG-4. The achievable bit rate for a five-channel coding is 320 kbit/s. In the following the main signal processing steps for MPEG-2 AAC are introduced and the principle functionalities explained. An extensive explanation can be found in [Bos97, Bra98, Bos02]. The MPEG-2 AAC coder is shown in Fig. 9.19. The corresponding decoder performs the functional units in reverse order with corresponding decoder functionalities.

Pre-processing. The input signal will be band-limited according to the sampling frequency. This step is used only in the *scalable sampling rate profile* [Bos97, Bra98, Bos02].

Filter bank. The time-frequency decomposition into $M = 1024$ subbands with an overlapped MDCT [Pri86, Pri87] is based on blocks of $N = 2048$ input samples. A stepwise explanation of the implementation is given. A graphical representation of the single steps is shown in Fig. 9.20. The single steps are as follows:

1. Partitioning of the input signal $x(n)$ with time index n into overlapped blocks

$$x_m(r) = x(mM + r), \quad r = 0, \dots, N - 1; -\infty \le m \le \infty, \qquad (9.33)$$

   of length N with an overlap (hop size) of $M = N/2$. The time index inside a block is denoted by r. The variable m denotes the block index.

2. Windowing of blocks with window function $w(r) \to x_m(r) \cdot w(r)$.

3. The MDCT

$$X(m, k) = \sqrt{\frac{2}{M}} \sum_{r=0}^{N-1} x_m(r) w(r) \cos\left(\frac{\pi}{M}\left(k + \frac{1}{2}\right)\left(r + \frac{M+1}{2}\right)\right),$$
$$k = 0, \dots, M - 1, \qquad (9.34)$$

   yields, for every M input samples, $M = N/2$ spectral coefficients from N windowed input samples.

4. Quantization of spectral coefficients $X(m, k)$ leads to quantized spectral coefficients $X_Q(m, k)$ based on a psychoacoustic model.

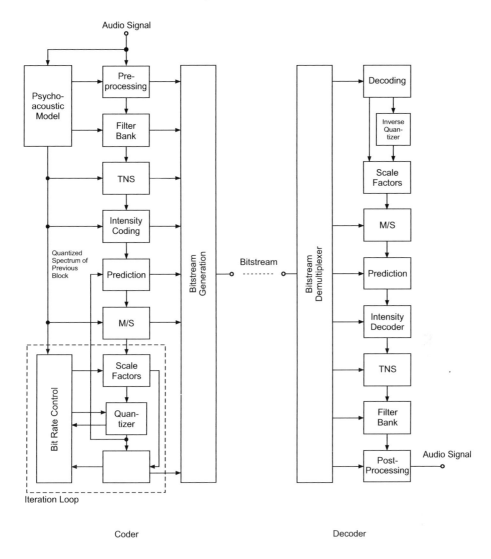

Figure 9.19 MPEG-2 AAC coder and decoder.

5. The IMDCT (Inverse Modified Discrete Cosine Transform)

$$\hat{x}_m(r) = \sqrt{\frac{2}{M}} \sum_{k=0}^{M-1} X_Q(m, k) \cos\left(\frac{\pi}{M}\left(k + \frac{1}{2}\right)\left(r + \frac{M+1}{2}\right)\right),$$
$$r = 0, \ldots, N-1, \tag{9.35}$$

yields, for every M input samples, N output samples in block $\hat{x}_m(r)$.

6. Windowing of inverse transformed block $\hat{x}_m(r)$ with window function $w(r)$.

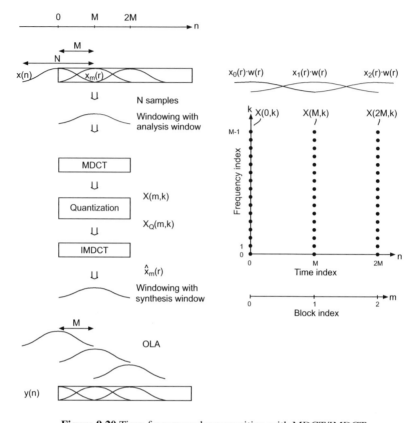

Figure 9.20 Time-frequency decomposition with MDCT/IMDCT.

7. Reconstruction of output signal $y(n)$ by overlap-add operation according to

$$y(n) = \sum_{m=-\infty}^{\infty} \hat{x}_m(r) w(r), \quad r = 0, \ldots, N-1, \tag{9.36}$$

with overlap M.

In order to explain the procedural steps we consider the MDCT/IMDCT of a sine pulse shown in Fig. 9.21. The left column shows from the top down the input signal and partitions of the input signal of block length $N = 256$. The window function is a sine window. The corresponding MDCT coefficients of length $M = 128$ are shown in the middle column. The IMDCT delivers the signals in the right column. One can observe that the inverse transforms with the IMDCT do not exactly reconstruct the single input blocks. Moreover, each output block consists of an input block and a special superposition of a time-reversed and by $M = N/2$ circular shifted input block, which is denoted by time-domain aliasing [Pri86, Pri87, Edl89]. The overlap-add operation of the single output blocks perfectly recovers the input signal which is shown in the top signal of the right column (Fig. 9.21). For a perfect reconstruction of the output signal, the window function of the analysis and synthesis step has to fulfill the condition $w^2(r) + w^2(r + M) = 1, r = 0, \ldots, M-1$.

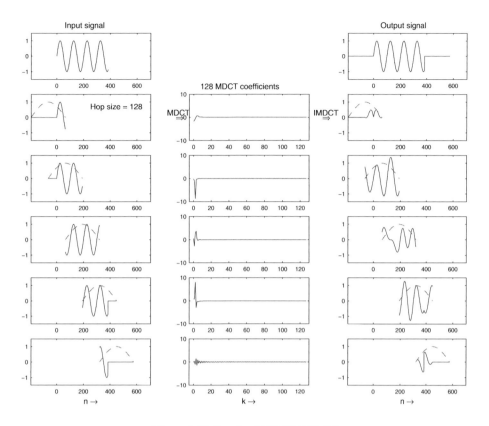

Figure 9.21 Signals of MDCT/IMDCT.

The Kaiser–Bessel derived window [Bos02] and a sine window $h(n) = \sin((n + \frac{1}{2})\frac{\pi}{N})$ with $n = 0, \ldots, N - 1$ [Mal92] are applied. Figure 9.22 shows both window functions with $N = 2048$ and the corresponding magnitude responses for a sampling frequency of $f_S = 44100$ Hz. The sine window has a smaller pass-band width but slower falling side lobes. In contrast, the Kaiser–Bessel derived window shows a wider pass-band and a faster decay of the side lobes. In order to demonstrate the filter bank properties and in particular the frequency decomposition of MDCT, we derive the modulated band-pass impulse responses of the window functions (prototype impulse response $w(n) = h(n)$) according to

$$h_k(n) = 2 \cdot h(n) \cdot \cos\left(\frac{\pi}{M}\left(k + \frac{1}{2}\right)\left(n + \frac{M+1}{2}\right)\right),$$
$$k = 0, \ldots, M - 1; \ n = 0, \ldots, N - 1. \qquad (9.37)$$

Figure 9.23 shows the normalized prototype impulse response of the sine window and the first two modulated band-pass impulse responses $h_0(n)$ and $h_1(n)$ and accordingly the corresponding magnitude responses are depicted. Besides the increased frequency resolution with $M = 1024$ band-pass filters, the reduced stop-band attenuation can be observed. A comparison of this magnitude response of the MDCT with the frequency resolution of

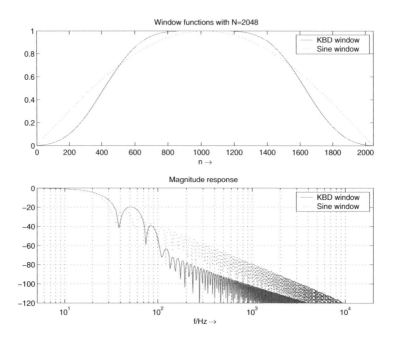

Figure 9.22 Kaiser–Bessel derived window and sine window for $N = 2048$ and magnitude responses of the normalized window functions.

the PQMF filter bank with $M = 32$ in Fig. 9.16 points out the different properties of both subband decompositions.

For adjusting the time and frequency resolution to the properties of an audio signal several methods have been investigated. Signal-adaptive audio coding based on the wavelet transform can be found in [Sin93, Ern00]. Window switching can be applied for achieving a time-variant time-frequency resolution for MDCT and IMDCT applications. For stationary signals a high frequency resolution and a low time resolution are necessary. This leads to long windows with $N = 2048$. Coding of attacks of instruments needs a high time resolution (reduction of window length to $N = 256$) and thus reduces frequency resolution (reduction of number of spectral coefficients). A detailed description of switching between time-frequency resolution with the MDCT/IMDCT can be found in [Edl89, Bos97, Bos02]. Examples of switching between different window functions and windows of different length are shown in Fig. 9.24.

Temporal Noise Shaping. A further method for adapting the time-frequency resolution of a filter bank and here an MDCT/IMDCT to the signal characteristic is based on linear prediction along the spectral coefficients in the frequency domain [Her96, Her99]. This method is called *temporal noise shaping* (TNS) and is a weighting of the temporal envelope of the time-domain signal. Weighting the temporal envelope in this way is demonstrated in Fig. 9.25.

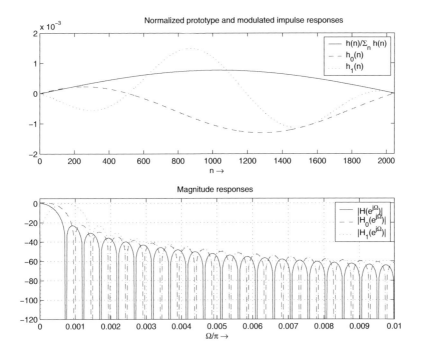

Figure 9.23 Normalized impulse responses of sine window for $N = 2048$, modulated band-pass impulse responses, and magnitude responses.

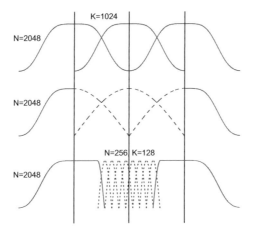

Figure 9.24 Switching of window functions.

Figure 9.25a shows a signal from a castanet attack. Making use of the discrete cosine transform (DCT, [Rao90])

$$X^{C(2)}(k) = \sqrt{\frac{2}{N}} c_k \sum_{n=0}^{N-1} x(n) \cos\left(\frac{(2n+1)k\pi}{2N}\right), \quad k = 0, \ldots, N-1 \quad (9.38)$$

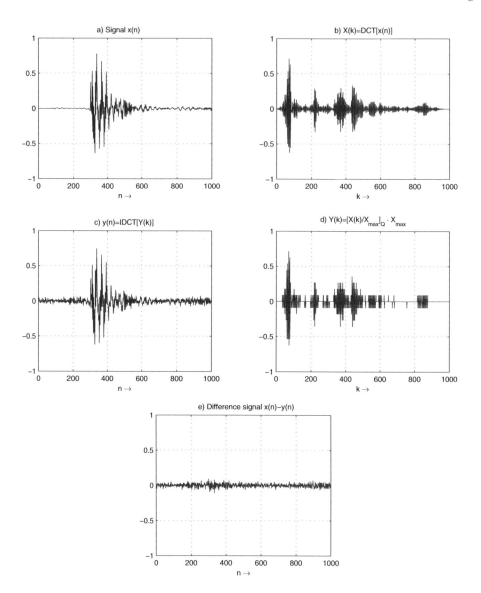

Figure 9.25 Attack of castanet and spectrum.

and the inverse discrete cosine transform (IDCT)

$$x(n) = \sqrt{\frac{2}{N}} \sum_{k=0}^{N-1} c_k X^{C(2)}(k) \cos\left(\frac{(2n+1)k\pi}{2N}\right), \quad n = 0, \ldots, N-1$$

$$\text{with } c_k = \begin{cases} 1/\sqrt{(2)}, & k = 0, \\ 1, & \text{otherwise,} \end{cases} \tag{9.39}$$

the spectral coefficients of the DCT of this castanet attack are represented in Fig. 9.25b. After quantization of these spectral coefficients $X(k)$ to 4 bits (Fig. 9.25d) and IDCT of the quantized spectral coefficients, the time-domain signal in Fig. 9.25c and the difference signal in Fig. 9.25e between input and output result. One can observe in the output and difference signal that the error is spread along the entire block length. This means that before the attack of the castanet happens, the error signal of the block is perceptible. The time-domain masking, referred to as pre-masking [Zwi90], is not sufficient. Ideally, the spreading of the error signal should follow the time-domain envelope of the signal itself. From forward linear prediction in the time domain it is known that the power spectral density of the error signal after coding and decoding is weighted by the envelope of the power spectral density of the input signal [Var06]. Performing a forward linear prediction along the frequency axis in the frequency domain and quantization and coding leads to an error signal in the time domain where the temporal envelope of the error signal follows the time-domain envelope of the input signal [Her96]. To point out the temporal weighting of the error signal we consider the forward prediction in the time domain in Fig. 9.26a. For coding the input signal, $x(n)$ is predicted by an impulse response $p(n)$. The output of the predictor is subtracted from the input signal $x(n)$ and delivers the signal $d(n)$, which is then quantized to a reduced word-length. The quantized signal $d_Q(n) = x(n) * a(n) + e(n)$ is the sum of the convolution of $x(n)$ with the impulse response $a(n)$ and the additive quantization error $e(n)$. The power spectral density of the coder output is $S_{D_Q D_Q}(e^{j\Omega}) = S_{XX}(e^{j\Omega}) \cdot |A(e^{j\Omega})|^2 + S_{EE}(e^{j\Omega})$. The decoding operation performs the convolution of $d_Q(n)$ with the impulse response $h(n)$ of the inverse system to the coder. Therefore $a(n) * h(n) = \delta(n)$ must hold and thus $H(e^{j\Omega}) = 1/A(e^{j\Omega})$. Hereby the output signal $y(n) = x(n) + e(n) * h(n)$ is derived with the corresponding discrete Fourier transform $Y(k) = X(k) + E(k) \cdot H(k)$. The power spectral density of the decoder out signal is given by $S_{YY}(e^{j\Omega}) = S_{XX}(e^{j\Omega}) + S_{EE}(e^{j\Omega}) \cdot |H(e^{j\Omega})|^2$. Here one can observe the spectral weighting of the quantization error with the spectral envelope of the input signal which is represented by $|H(e^{j\Omega})|$. The same kind of forward prediction will now be applied in the frequency domain to the spectral coefficients $X(k) = \text{DCT}[x(n)]$ for a block of input samples $x(n)$ shown in Fig. 9.26b. The output of the decoder is then given by $Y(k) = X(k) + E(k) * H(k)$ with $A(k) * H(k) = \delta(k)$. Thus, the corresponding time-domain signal is $y(n) = x(n) + e(n) \cdot h(n)$, where the temporal weighting of the quantization error with the temporal envelope of the input signal is clearly evident. The temporal envelope is represented by the absolute value $|h(n)|$ of the impulse response $h(n)$. The relation between the temporal signal envelope (absolute value of the analytical signal) and the autocorrelation function of the analytical spectrum is discussed in [Her96]. The dualities between forward linear prediction in time and frequency domain are summarized in Table 9.2. Figure 9.27 demonstrates the operations for temporal noise shaping in the coder, where the prediction is performed along the spectral coefficients. The coefficients of the forward predictor have to be transmitted to the decoder, where the inverse filtering is performed along the spectral coefficients.

The temporal weighting is finally demonstrated in Fig. 9.28, where the corresponding signals with forward prediction in the frequency domain are shown. Figure 9.28a,b shows the castanet signal $x(n)$ and its corresponding spectral coefficients $X(k)$ of the applied DCT. The forward prediction delivers $D(k)$ in Fig. 9.28d and the quantized signal $D_Q(k)$ in Fig. 9.28f. After the decoder the signal $Y(k)$ in Fig. 9.28h is reconstructed by the inverse

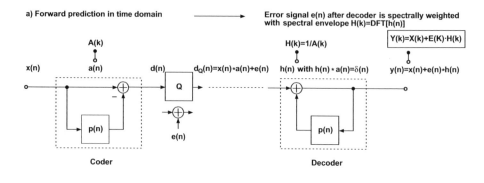

a) Forward prediction in time domain $\longrightarrow$ Error signal e(n) after decoder is spectrally weighted with spectral envelope H(k)=DFT[h(n)]

Coder   Decoder

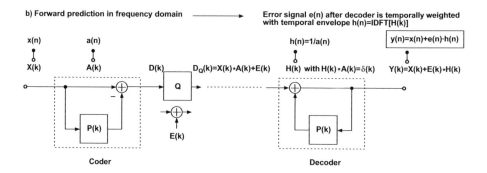

b) Forward prediction in frequency domain $\longrightarrow$ Error signal e(n) after decoder is temporally weighted with temporal envelope h(n)=IDFT[H(k)]

Coder   Decoder

Figure 9.26 Forward prediction in time and frequency domain.

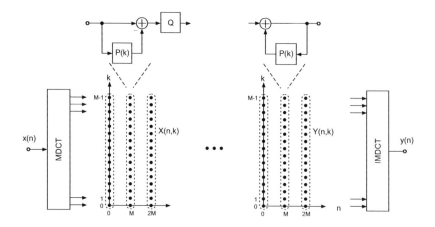

Figure 9.27 Temporal noise shaping with forward prediction in frequency domain.

transfer function. The IDCT of $Y(k)$ finally results in the output signal $y(n)$ in Fig. 9.28e. The difference signal $x(n) - y(n)$ in Fig. 9.28g demonstrates the temporal weighting of the error signal with the temporal envelope from Fig. 9.28c. For this example, the order of the

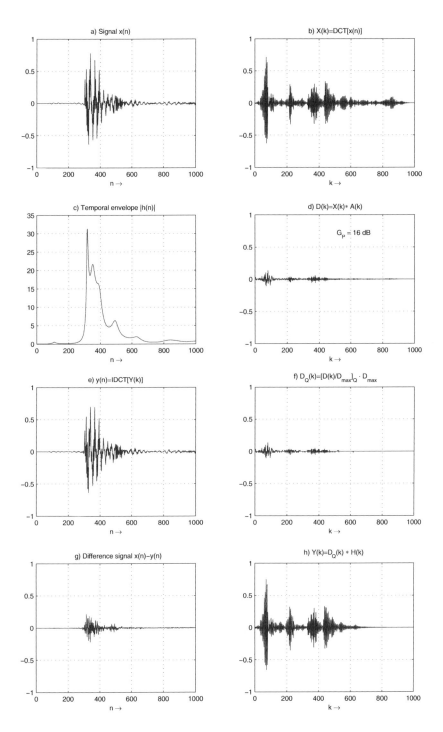

Figure 9.28 Temporal noise shaping: attack of castanet and spectrum.

Table 9.2 Forward prediction in time and frequency domain.

Prediction in time domain	Prediction in frequency domain
$y(n) = x(n) + e(n) * h(n)$	$y(n) = x(n) + e(n) \cdot h(n)$
$Y(k) = X(k) + E(k) \cdot H(k)$	$Y(k) = X(k) + E(k) * H(k)$

predictor is chosen as 20 [Bos97] and the prediction along the spectral coefficients $X(k)$ is performed by the Burg method. The prediction gain for this signal in the frequency domain is $G_p = 16$ dB (see Fig. 9.28d).

Frequency-domain Prediction. A further compression of the band-pass signals is possible by using linear prediction. A backward prediction [Var06] of the band-pass signals is applied on the coder side (see Fig. 9.29). In using a backward prediction the predictor coefficients need not be coded and transmitted to the decoder, since the estimate of the input sample is based on the quantized signal. The decoder derives the predictor coefficients $p(n)$ in the same way from the quantized input. A second-order predictor is sufficient, because the bandwidth of the band-pass signals is very low [Bos97].

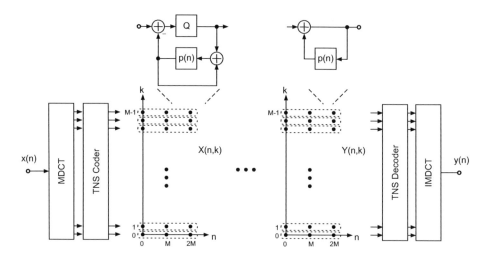

Figure 9.29 Backward prediction of band-pass signals.

Mono/Side Coding. Coding of stereo signals with left and right signals $x_L(n)$ and $x_R(n)$ can be achieved by coding a mono signal (M) $x_M(n) = (x_L(n) + x_R(n))/2$ and a side (S, difference) signal $x_S(n) = (x_L(n) - x_R(n))/2$ (M/S coding). Since for highly correlated left and right signals the power of the side signal is reduced, a reduction in bit rate for this signal can be achieved. The decoder can reconstruct the left signal $x_L(n) = x_M(n) + x_S(n)$ and the right signal $x_R(n) = x_M(n) - x_S(n)$, if no quantization and coding is applied to the mono and side signal. This M/S coding is carried out for MPEG-2 AAC [Bra98, Bos02] with the spectral coefficients of a stereo signal (see Fig. 9.30).

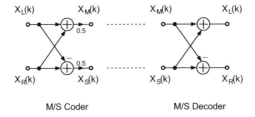

Figure 9.30 M/S coding in frequency domain.

Intensity Stereo Coding. For intensity stereo (IS) coding a mono signal $x_M(n) = x_L(n) + x_R(n)$ and two temporal envelopes $e_L(n)$ and $e_R(n)$ of the left and right signals are coded and transmitted. On the decoding side the left signal is reconstructed by $y_L(n) = x_M(n) \cdot e_L(n)$ and the right signal by $y_R(n) = x_M(n) \cdot e_R(n)$. This reconstruction is lossy. The IS coding of MPEG-2 AAC [Bra98] is performed by summation of spectral coefficients of both signals and by coding of scale factors which represent the temporal envelope of both signals (see Fig. 9.31). This type of stereo coding is only useful for higher frequency bands, since the human perception for phase shifts is non-sensitive for frequencies above 2 kHz.

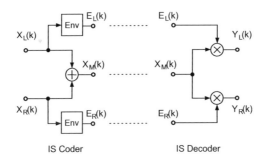

Figure 9.31 Intensity stereo coding in frequency domain.

Quantization and Coding. During the last coding step the quantization and coding of the spectral coefficients takes place. The quantizers, which are used in the figures for prediction along spectral coefficients in frequency direction (Fig. 9.27) and prediction in the frequency domain along band-pass signals (Fig. 9.29), are now combined into a single quantizer per spectral coefficient. This quantizer performs nonlinear quantization similar to a floating-point quantizer of Chapter 2 such that a nearly constant signal-to-noise ratio over a wide amplitude range is achieved. This floating-point quantization with a so-called scale factor is applied to several frequency bands, in which several spectral coefficients use a common scale factor derived from an iteration loop (see Fig. 9.19). Finally, a Huffman coding of the quantized spectral coefficients is performed. An extensive presentation can be found in [Bos97, Bra98, Bos02].

9.7 MPEG-4 Audio Coding

The MPEG-4 audio coding standard consists of a family of audio and speech coding methods for different bit rates and a variety of multimedia applications [Bos02, Her02]. Besides a higher coding efficiency, new functionalities such as scalability, object-oriented representation of signals and interactive synthesis of signals at the decoder are integrated. The MPEG-4 coding standard is based on the following speech and audio coders.

- Speech coders

    - CELP: Code Excited Linear Prediction (bit rate 4–24 kbit/s).
    - HVXC: Harmonic Vector Excitation Coding (bit rate 1.4–4 kbit/s).

- Audio coders

    - Parametric audio: representation of a signal as a sum of sinsoids, harmonic components, and residual components (bit rate 4–16 kbit/s).
    - Structured audio: synthetic signal generation at decoder (extension of the MIDI standard[1]) (200 bit–4 kbit/s).
    - Generalized audio: extension of MPEG-2 AAC with additional methods in the time-frequency domain. The basic structure is depicted in Fig. 9.19 (bit rate 6–64 kbit/s).

Basics of speech coders can found in [Var06]. The specified audio coders allow coding with lower bit rates (Parametric Audio and Structured Audio) and coding with higher quality at lower bit rates compared to MPEG-2 AAC.

Compared to coding methods such as MPEG-1 and MPEG-2 introduced in previous sections, the parametric audio coding is of special interest as an extension to the filter bank methods [Pur99, Edl00]. A parametric audio coder is shown in Fig. 9.32. The analysis of the audio signal leads to a decomposition into sinusoidal, harmonic and noise-like signal components and the quantization and coding of these signal components is based on psychoacoustics [Pur02a]. According to an analysis/synthesis approach [McA86, Ser89, Smi90, Geo92, Geo97, Rod97, Mar00a] shown in Fig. 9.33 the audio signal is represented in a parametric form given by

$$x(n) = \sum_{i=1}^{M} A_i(n) \cos\left(2\pi \frac{f_i(n)}{f_A} n + \varphi_i(n)\right) + x_n(n). \tag{9.40}$$

The first term describes a sum of sinusoids with time-varying amplitudes $A_i(n)$, frequencies $f_i(n)$ and phases $\varphi_i(n)$. The second term consists of a noise-like component $x_n(n)$ with time-varying temporal envelope. This noise-like component $x_n(n)$ is derived by subtracting the synthesized sinusoidal components from the input signal. With the help of a further analysis step, harmonic components with a fundamental frequency and multiples of this fundamental frequency are identified and grouped into harmonic components. The extraction of deterministic and stochastic components from an audio signal can be found in

---

[1] http://www.midi.org/.

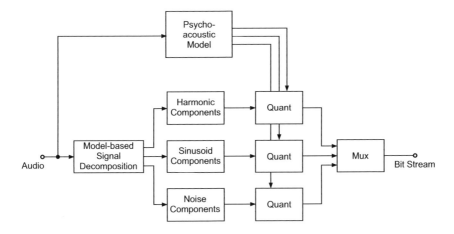

Figure 9.32 MPEG-4 parametric coder.

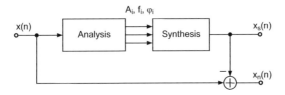

Figure 9.33 Parameter extraction with analysis/synthesis.

[Alt99, Hai03, Kei01, Kei02, Mar00a, Mar00b, Lag02, Lev98, Lev99, Pur02b]. In addition to the extraction of sinusoidal components, the modeling of noise-like components and transient components is of specific importance [Lev98, Lev99]. Figure 9.34 exemplifies the decomposition of an audio signal into a sum of sinusoids $x_s(n)$ and a noise-like signal $x_n(n)$. The spectrogram shown in Fig. 9.35 represents the short-time spectra of the sinusoidal components. The extraction of the sinusoids has been achieved by a modified FFT method [Mar00a] with an FFT length of $N = 2048$ and an analysis hop size of $R_A = 512$.

The corresponding parametric MPEG-4 decoder is shown in Fig. 9.36 [Edl00, Mei02]. The synthesis of the three signal components can be achieved by inverse FFT and overlap-add methods or can be directly performed by time-domain methods [Rod97, Mei02]. A significant advantage of parametric audio coding is the direct access at the decoder to the three main signal components which allows effective post-processing for the generation of a variety of audio effects [Zöl02]. Effects such as time and pitch scaling, virtual sources in three-dimensional spaces and cross-synthesis of signals (karaoke) are just a few examples of interactive sound design on the decoding side.

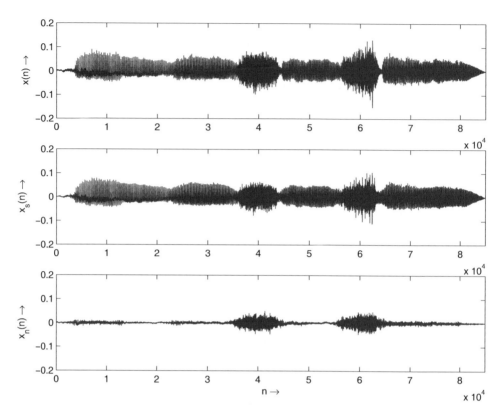

Figure 9.34 Original signal, sum of sinusoids and noise-like signal.

9.8   Spectral Band Replication

To further reduce the bit rate an extension of MPEG-1 Layer III with the name MP3pro was introduced [Die02, Zie02]. The underlying method, called *spectral band replication* (SBR), performs a low-pass and high-pass decomposition of the audio signal, where the low-pass filtered part is coded by a standard coding method (e.g. MPEG-1 Layer III) and the high-pass part is represented by a spectral envelope and a difference signal [Eks02, Zie03]. Figure 9.37 shows the functional units of an SBR coder. For the analysis of the difference signal the high-pass part (HP Generator) is reconstructed from the low-pass part and compared to the actual high-pass part. The difference is coded and transmitted. For decoding (see Fig. 9.38) the decoded low-pass part of a standard decoder is used by the HP generator to reconstruct the high-pass part. The additional coded difference signal is added at the decoder. An equalizer provides the spectral envelope shaping for the high-pass part. The spectral envelope of the high-pass signal can be achieved by a filter bank and computing the RMS values of each band-pass signal [Eks02, Zie03]. The reconstruction of the high-pass part (HP Generator) can also be achieved by a filter bank and substituting the band-pass signals by using the low-pass parts [Schu96, Her98]. To code the difference

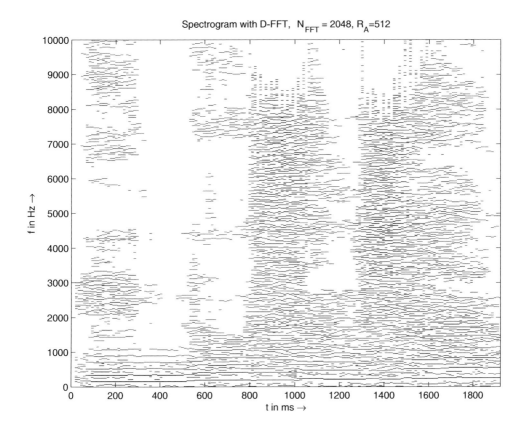

Figure 9.35 Spectrogram of sinusoidal components.

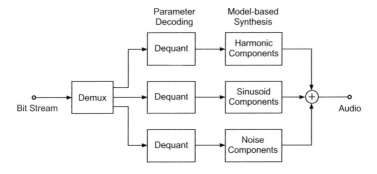

Figure 9.36 MPEG-4 parametric decoder.

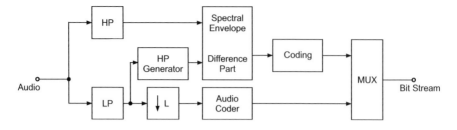

Figure 9.37 SBR coder.

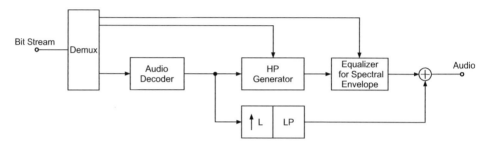

Figure 9.38 SBR decoder.

signal of the high-pass part additive sinusoidal models can be applied such as the parametric methods of the MPEG-4 coding approach.

Figure 9.39 shows the functional units of the SBR method in the frequency domain. First, the short-time spectrum is used to calculate the spectral envelope (Fig. 9.39a). The spectral envelope can be derived from an FFT, a filter bank, the cepstrum or by linear prediction [Zöl02]. The band-limited low-pass signal can be downsampled and coded by a standard coder which operates at a reduced sampling rate. In addition, the spectral envelope has to be coded (Fig. 9.39b). On the decoding side the reconstruction of the upper spectrum is achieved by frequency-shifting of the low-pass part or even specific low-pass parts and applying the spectral envelope onto this artificial high-pass spectrum (Fig. 9.39c). An efficient implementation of a time-varying spectral envelope computation (at the coder side) and spectral weighting of the high-pass signal (at the decoder side) with a complex-valued QMF filter bank is described in [Eks02].

9.9 Java Applet – Psychoacoustics

The applet shown in Fig. 9.40 demonstrates psychoacoustic audio masking effects [Gui05]. It is designed for a first insight into the perceptual experience of masking a sinusoidal signal with band-limited noise.

You can choose between two predefined audio files from our web server (*audio1.wav* or *audio2.wav*). These are band-limited noise signals with different frequency ranges. A sinusoidal signal is generated by the applet, and two sliders can be used to control its frequency and magnitude values.

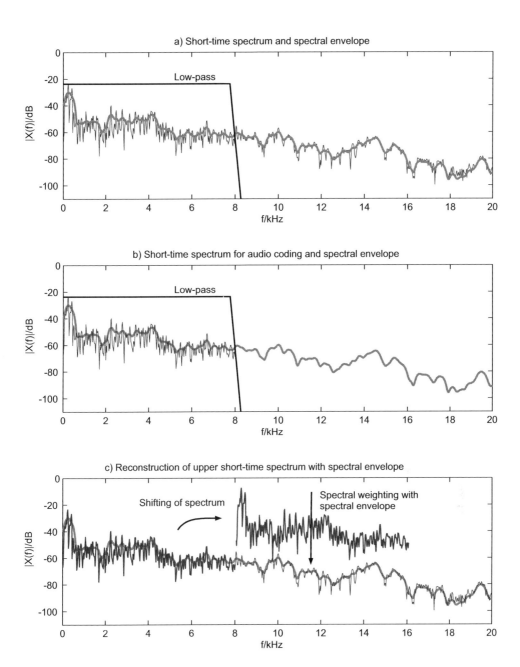

Figure 9.39 Functional units of SBR method.

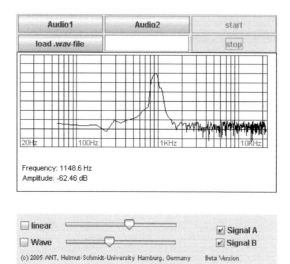

Figure 9.40 Java applet – psychoacoustics.

9.10 Exercises

1. Psychoacoustics

1. Human hearing

   (a) What is the frequency range of human sound perception?

   (b) What is the frequency range of speech?

   (c) In the above specified range where is the human hearing most sensitive?

   (d) Explain how the absolute threshold of hearing has been obtained.

2. Masking

   (a) What is frequency-domain masking?

   (b) What is a critical band and why is it needed for frequency masking phenomena?

   (c) Consider a_i and f_i to be respectively the amplitude and the frequency of a partial at index i and $V(a_i)$ to be the corresponding volume in dB. The difference between the level of the masker and the masking threshold is -10 dB. The masking curves toward lower and higher frequencies are described respectively by a left slope (27 dB/Bark) and a right slope (15 dB/Bark). Explain the main steps of frequency masking in this case and show with plots how this masking phenomena is achieved.

   (d) What are the psychoacoustic parameters used for lossy audio coding?

   (e) How can we explain the temporal masking and what is its duration after stopping the active masker?

2. Audio coding

1. Explain the lossless coder and decoder.

2. What is the achievable compression factor for lossless coding?

3. Explain the MPEG-1 Layer III coder and decoder.

4. Explain the MPEG-2 AAC coder and decoder.

5. What is temporal noise shaping?

6. Explain the MPEG-4 coder and decoder.

7. What is the benefit of SBR?

References

[Alt99] R. Althoff, F. Keiler, U. Zölzer: *Extracting Sinusoids from Harmonic Signals*, Proc. DAFX-99 Workshop on Digital Audio Effects, pp. 97–100, Trondheim, 1999.

[Blo95] T. Block: *Untersuchung von Verfahren zur verlustlosen Datenkompression von digitalen Audiosignalen*, Studienarbeit, TU Hamburg-Harburg, 1995.

[Bos97] M. Bosi, K. Brandenburg, S. Quackenbush, L. Fielder, K. Akagiri, H. Fuchs, M. Dietz, J. Herre, G. Davidson, Y. Oikawa: *ISO/IEC MPEG-2 Advanced Audio Coding*, J. Audio Eng. Soc., Vol. 45, pp. 789–814, October 1997.

[Bos02] M. Bosi, R. E. Goldberg: *Introduction to Digital Audio Coding and Standards*, Kluwer Academic, Boston, 2002.

[Bra92] K. Brandenburg, J. Herre: *Digital Audio Compression for Professional Applications*, Proc. 92nd AES Convention, Preprint No. 3330, Vienna, 1992.

[Bra94] K. Brandenburg, G. Stoll: *The ISO-MPEG-1 Audio: A Generic Standard for Coding of High Quality Digital Audio*, J. Audio Eng. Soc., Vol. 42, pp. 780–792, October 1994.

[Bra98] K. Brandenburg: *Perceptual Coding of High Quality Digital Audio*, in M. Kahrs, K. Brandenburg (Ed.), Applications of Digital Signal Processing to Audio and Acoustics, Kluwer Academic, Boston, 1998.

[Cel93] C. Cellier, P. Chenes, M. Rossi: *Lossless Audio Data Compression for Real-Time Applications*, Proc. 95th AES Convention, Preprint No. 3780, New York, 1993.

[Cra96] P. Craven, M. Gerzon: *Lossless Coding for Audio Discs*, J. Audio Eng. Soc., Vol. 44, pp. 706–720, 1996.

[Cra97]   P. Craven, M. Law, J. Stuart: *Lossless Compression Using IIR Prediction*, Proc. 102nd AES Convention, Preprint No. 4415, Munich, 1997.

[Die02]   M. Dietz, L. Liljeryd, K. Kjörling, O. Kunz: *Spectral Band Replication: A Novel Approach in Audio Coding*, Proc. 112th AES Convention, Preprint No. 5553, Munich, 2002.

[Edl89]   B. Edler: *Codierung von Audiosignalen mit überlappender Transformation und adaptiven Fensterfunktionen*, Frequenz, Vol. 43, pp. 252–256, 1989.

[Edl00]   B. Edler, H. Purnhagen: *Parametric Audio Coding*, 5th International Conference on Signal Processing (ICSP 2000), Beijing, August 2000.

[Edl95]   B. Edler: *Äquivalenz von Transformation und Teilbandzerlegung in der Quellencodierung*, Dissertation, Universität Hannover, 1995.

[Eks02]   P. Ekstrand: *Bandwidth Extension of Audio Signals by Spectral Band Replication*, Proc. 1st IEEE Benelux Workshop on Model-Based Processing and Coding of Audio (MPCA-2002), Leuven, Belgium, 2002.

[Ern00]   M. Erne: *Signal Adpative Audio Coding Using Wavelets and Rate Optimization*, Dissertation, ETH Zurich, 2000.

[Geo92]   E. B. George, M. J. T. Smith: *Analysis-by-Synthesis/Overlap-Add Sinusoidal Modeling Applied to the Analysis and Synthesis of Musical Tones*, J. Audio Eng. Soc., Vol. 40, pp. 497–516, June 1992.

[Geo97]   E. B. George, M. J. T. Smith: *Speech Analysis/Synthesis and Modification using an Analysis-by-Synthesis/Overlap-Add Sinusoidal Model*, IEEE Trans. on Speech and Audio Processing, Vol. 5, No. 5, pp. 389–406, September 1997.

[Glu93]   R. Gluth: *Beiträge zur Beschreibung und Realisierung digitaler, nichtrekursiver Filterbänke auf der Grundlage linearer diskreter Transformationen*, Dissertation, Ruhr-Universität Bochum, 1993.

[Gui05]   M. Guillemard, C. Ruwwe, U. Zölzer: *J-DAFx – Digital Audio Effects in Java*, Proc. 8th Int. Conference on Digital Audio Effects (DAFx-05), pp. 161–166, Madrid, 2005.

[Hai03]   S. Hainsworth, M. Macleod: *On Sinusoidal Parameter Estimation*, Proc. DAFX-03 Conference on Digital Audio Effects, London, September 2003.

[Han98]   M. Hans: *Optimization of Digital Audio for Internet Transmission*, PhD thesis, Georgia Inst. Technol., Atlanta, 1998.

[Han01]   M. Hans, R. W. Schafer: *Lossless Compression of Digital Audio*, IEEE Signal Processing Magazine, Vol. 18, No. 4, pp. 21–32, July 2001.

[Hel72]   R. P. Hellman: *Asymmetry in Masking between Noise and Tone*, Perception and Psychophys., Vol. 11, pp. 241–246, 1972.

[Her96]   J. Herre, J. D. Johnston: *Enhancing the Performance of Perceptual Audio Coders by Using Temporal Noise Shaping (TNS)*, Proc. 101st AES Convention, Preprint No. 4384, Los Angeles, 1996.

[Her98]   J. Herre, D. Schultz: *Extending the MPEG-4 AAC Codec by Perceptual Noise Substitution*, Proc. 104th AES Convention, Preprint No. 4720, Amsterdam, 1998.

[Her99]   J. Herre: *Temporal Noise Shaping, Quantization and Coding Methods in Perceptual Audio Coding: A Tutorial Introduction*, Proc. AES 17th International Conference on High Quality Audio Coding, Florence, September 1999.

[Her02]   J. Herre, B. Grill: *Overview of MPEG-4 Audio and its Applications in Mobile Communications*, Proc. 112th AES Convention, Preprint No. 5553, Munich, 2002.

[Huf52]   D. A. Huffman: *A Method for the Construction of Minimum-Redundancy Codes*, Proc. of the IRE, Vol. 40, pp. 1098–1101, 1952.

[ISO92]   ISO/IEC 11172-3: *Coding of Moving Pictures and Associated Audio for Digital Storage Media at up to 1.5 Mbits/s – Audio Part*, International Standard, 1992.

[Jay84]   N. S. Jayant, P. Noll: *Digital Coding of Waveforms*, Prentice Hall, Englewood Cliffs, NJ, 1984.

[Joh88a]   J. D. Johnston: *Transform Coding of Audio Signals Using Perceptual Noise Criteria*, IEEE J. Selected Areas in Communications, Vol. 6, No. 2, pp. 314–323, February 1988.

[Joh88b]   J. D. Johnston: *Estimation of Perceptual Entropy Using Noise Masking Criteria*, Proc. ICASSP-88, pp. 2524–2527, 1988.

[Kap92]   R. Kapust: *A Human Ear Related Objective Measurement Technique Yields Audible Error and Error Margin*, Proc. 11th Int. AES Conference – Test & Measurement, Portland, pp. 191–202, 1992.

[Kei01]   F. Keiler, U. Zölzer: *Extracting Sinusoids from Harmonic Signals*, J. New Music Research, Special Issue: Musical Applications of Digital Signal Processing, Guest Editor: Mark Sandler, Vol. 30, No. 3, pp. 243–258, September 2001.

[Kei02]   F. Keiler, S. Marchand: *Survey on Extraction of Sinusoids in Stationary Sounds*, Proc. DAFX-02 Conference on Digital Audio Effects, pp. 51–58, Hamburg, 2002.

[Kon94]   K. Konstantinides: *Fast Subband Filtering in MPEG Audio Coding*, IEEE Signal Processing Letters, Vol. 1, No. 2, pp. 26–28, February 1994.

[Lag02]   M. Lagrange, S. Marchand, J.-B. Rault: *Sinusoidal Parameter Extraction and Component Selection in a Non Stationary Model*, Proc. DAFX-02 Conference on Digital Audio Effects, pp. 59–64, Hamburg, 2002.

[Lev98]   S. Levine: *Audio Representations for Data Compression and Compressed Domain Processing*, PhD thesis, Stanford University, 1998.

[Lev99]   S. Levine, J. O. Smith: *Improvements to the Switched Parametric & Transform Audio Coder*, Proc. 1999 IEEE Workshop on Applications of Signal Processing to Audio and Acoustics, New Paltz, NY, October 1999.

[Lie02]   T. Liebchen: *Lossless Audio Coding Using Adaptive Multichannel Prediction*, Proc. 113th AES Convention, Preprint No. 5680, Los Angeles, 2002.

[Mar00a]   S. Marchand: *Sound Models for Computer Music*, PhD thesis, University of Bordeaux, October 2000.

[Mar00b]   S. Marchand: *Compression of Sinusoidal Modeling Parameters*, Proc. DAFX-00 Conference on Digital Audio Effects, pp. 273–276, Verona, December 2000.

[Mal92]   H. S. Malvar: *Signal Processing with Lapped Transforms*, Artech House, Boston, 1992.

[McA86]   R. McAulay, T. Quatieri: *Speech Transformations Based in a Sinusoidal Representation*, IEEE Trans. Acoustics, Speech, Signal Processing, Vol. 34, No. 4, pp. 744–754, 1989.

[Mas85]   J. Masson, Z. Picel: *Flexible Design of Computationally Efficient Nearly Perfect QMF Filter Banks*, Proc. ICASSP-85, pp. 541–544, 1985.

[Mei02]   N. Meine, H. Purnhagen: *Fast Sinusoid Synthesis For MPEG-4 HILN Parametric Audio Decoding*, Proc. DAFX-02 Conference on Digital Audio Effects, pp. 239–244, Hamburg, September 2002.

[Pen93]   W. B. Pennebaker, J. L. Mitchell: *JPEG Still Image Data Compression Standard*, Van Nostrand Reinhold, New York, 1993.

[Pri86]   J. P. Princen, A. B. Bradley: *Analysis/Synthesis Filter Bank Design Based on Time Domain Aliasing Cancellation*, IEEE Trans. on Acoustics, Speech, and Signal Processing, Vol. 34, No. 5, pp. 1153–1161, October 1986.

[Pri87]   J. P. Princen, A. W. Johnston, A. B. Bradley: *Subband/Transform Coding Using Filter Bank Designs Based on Time Domain Aliasing Cancellation*, Proc. ICASSP-87, pp. 2161–2164, 1987.

[Pur97]   M. Purat, T. Liebchen, P. Noll: *Lossless Transform Coding of Audio Signals*, Proc. 102nd AES Convention, Preprint No. 4414, Munich, 1997.

[Pur99]   H. Purnhagen: *Advances in Parametric Audio Coding*, Proc. 1999 IEEE Workshop on Applications of Signal Processing to Audio and Acoustics, New Paltz, NY, October 1999.

[Pur02a] H. Purnhagen, N. Meine, B. Edler: *Sinusoidal Coding Using Loudness-Based Component Selection*, Proc. ICASSP-2002, May 13–17, Orlando, FL, 2002.

[Pur02b] H. Purnhagen: *Parameter Estimation and Tracking for Time-Varying Sinusoids*, Proc. 1st IEEE Benelux Workshop on Model Based Processing and Coding of Audio, Leuven, Belgium, November 2002.

[Raa02] M. Raad, A. Mertins: *From Lossy to Lossless Audio Coding Using SPIHT*, Proc. DAFX-02 Conference on Digital Audio Effects, pp. 245–250, Hamburg, 2002.

[Rao90] K. R. Rao, P. Yip: *Discrete Cosine Transform – Algorithms, Advantages, Applications*, Academic Press, San Diego, 1990.

[Rob94] T. Robinson: *SHORTEN: Simple Lossless and Near-Lossless Waveform Compression*, Technical Report CUED/F-INFENG/TR.156, Cambridge University Engineering Department, Cambridge, December 1994.

[Rod97] X. Rodet: *Musical Sound Signals Analysis/Synthesis: Sinusoidal+Residual and Elementary Waveform Models*, Proceedings of the IEEE Time-Frequency and Time-Scale Workshop (TFTS-97), University of Warwick, Coventry, August 1997.

[Rot83] J. H. Rothweiler: *Polyphase Quadrature Filters – A New Subband Coding Technique*, Proc. ICASSP-87, pp. 1280–1283, 1983.

[Sauv90] U. Sauvagerd: *Bitratenreduktion hochwertiger Musiksignale unter Verwendung von Wellendigitalfiltern*, VDI-Verlag, Düsseldorf, 1990.

[Schr79] M. R. Schroeder, B. S. Atal, J. L. Hall: *Optimizing Digital Speech Coders by Exploiting Masking Properties of the Human Ear*, J. Acoust. Soc. Am., Vol. 66, No. 6, pp. 1647–1652, December 1979.

[Sch02] G. D. T. Schuller, Bin Yu, Dawei Huang, B. Edler: *Perceptual Audio Coding Using Adaptive Pre- and Post-Filters and Lossless Compression*, IEEE Trans. on Speech and Audio Processing, Vol. 10, No. 6, pp. 379–390, September 2002.

[Schu96] D. Schulz: *Improving Audio Codecs by Noise Substitution*, J. Audio Eng. Soc., Vol. 44, pp. 593–598, July/August 1996.

[Ser89] X. Serra: *A System for Sound Analysis/Transformation/Synthesis Based on a Deterministic plus Stochastic Decomposition*, PhD thesis, Stanford University, 1989.

[Sin93] D. Sinha, A. H. Tewfik: *Low Bit Rate Transparent Audio Compression Using Adapted Wavelets*, IEEE Trans. on Signal Processing, Vol. 41, pp. 3463–3479, 1993.

[Smi90] J. O. Smith, X. Serra: *Spectral Modeling Synthesis: A Sound Analysis/Synthesis System Based on a Deterministic plus Stochastic Decomposition*, Computer Music J., Vol. 14, No. 4, pp. 12–24, 1990.

[Sqa88]   EBU-SQAM: *Sound Quality Assessment Material*, Recordings for Subjective
           Tests, CompactDisc, 1988.

[Ter79]   E. Terhardt: *Calculating Virtual Pitch*, Hearing Res., Vol. 1, pp. 155–182, 1979.

[Thei88]   G. Theile, G. Stoll, M. Link: *Low Bit-Rate Coding of High-Quality Audio
           Signals*, EBU Review, No. 230, pp. 158–181, August 1988.

[Vai93]   P. P. Vaidyanathan: *Multirate Systems and Filter Banks*, Prentice Hall,
           Englewood Cliffs, NJ, 1993.

[Var06]   P. Vary, R. Martin: *Digital Speech Transmission. Enhancement, Coding and
           Error Concealment*, John Wiley & Sons, Ltd, Chichester, 2006.

[Vet95]   M. Vetterli, J. Kovacevic: *Wavelets and Subband Coding*, Prentice Hall,
           Englewood Cliffs, NJ, 1995.

[Zie02]   T. Ziegler, A. Ehret, P. Ekstrand, M. Lutzky: *Enhancing mp3 with SBR:
           Features and Capabilities of the new mp3PRO Algorithm*, Proc. 112th AES
           Convention, Preprint No. 5560, Munich, 2002.

[Zie03]   T. Ziegler, M. Dietz, K. Kjörling, A. Ehret: *aacPlus-Full Bandwidth Audio
           Coding for Broadcast and Mobile Applications*, International Signal Processing
           Conference, Dallas, 2003.

[Zöl02]   U. Zölzer (Ed.): *DAFX – Digital Audio Effects*, John Wiley & Sons, Ltd,
           Chichester, 2002.

[Zwi82]   E. Zwicker: *Psychoakustik*, Springer-Verlag, Berlin, 1982.

[Zwi90]   E. Zwicker, H. Fastl: *Psychoacoustics*, Springer-Verlag, Berlin, 1990.

Index

Printed and bound in the UK by
CPI Antony Rowe, Eastbourne